Reframing Economics

Reframing Economics

Economic Action as Imperfect Cooperation

Roger A. McCain

Professor of Economics, Drexel University, USA

Edward Elgar

Cheltenham, UK • Northampton, MA, USA

Published by
Edward Elgar Publishing Limited
The Lypiatts
15 Lansdown Road
Cheltenham
Glos GL50 2JA
UK

Edward Elgar Publishing, Inc.
William Pratt House
9 Dewey Court
Northampton
Massachusetts 01060
USA

A catalogue record for this book
is available from the British Library

Library of Congress Control Number: 2013954353

This book is available electronically in the ElgarOnline.com
Economics Subject Collection, E-ISBN 978 1 78254 642 9

ISBN 978 1 78254 641 2 (cased)

Typeset by Servis Filmsetting Ltd, Stockport, Cheshire
Printed and bound in Great Britain by T.J. International Ltd, Padstow

Contents

1. Introduction

Economics is an important study. Most would agree that it is crucial for business, for politics, and for many other aspects of our life. How, then, should we define economics?* It may come as a surprise that there is little consensus on this![1] Perhaps it doesn't matter very much. Economist Kenneth Boulding said that "Economics is what economists do." (He attributed the remark to his teacher Jacob Viner.) John Stuart Mill, one of the greatest economist-philosophers of the nineteenth century, said that it is best not to begin with a definition, but to end with a definition, after the reader has enough examples so that the definition is meaningful.[2] But we shall not follow his advice. A definition is a brief statement of what psychologists call a frame.[3] A frame or frame of reference in this sense is a way of posing and understanding a decision, and psychologists have shown that the way we frame a decision influences the decision we make and the actions we undertake. And as we approach the study of economics, we have to make decisions – decisions about what is more important, what is less important, where the efforts of researchers would be best focused, what can safely be left behind and neglected. These decisions will depend on our frame of reference, and while the frame of reference is always too complicated to be completely expressed as a definition, the definition is an important tool with which we begin to frame our study of economics.

INTERDEPENDENCE AND AFFLUENCE

Even if there is little consensus about the definition of economics, there is one definition that was particularly influential in the twentieth century, and it was stated by Lord Lionel Robbins:[4] "Economics is a science which studies human behaviour as a relationship between ends and scarce means which have alternative uses." By contrast, for Adam Smith it was (to quote the title of his founding book originally published in 1776) "an inquiry into the nature and causes of the wealth of nations." It seems clear that the efficient use of scarce resources (stressed by Robbins) is

* For this book superscript numbers in parentheses refer to the notes in the final section of each chapter, "Sources and Reading."

one important cause of the wealth of nations, but perhaps not the most important one and certainly not the only one. In that sense Robbins's definition is narrower than Smith's. Robbins's definition also stressed the separation of the economics of his time from that of the earlier period. His definition reflected a narrowing of the attention of economists that took place around 1900. Since then, however, economists' attention has again broadened, with more attention, for example, to innovation and information as aspects of economics. Perhaps the variety of definitions is in part a result of that broadening.

How, then, should we frame the study of economics? What links the current research in economics to the ideas of Smith and the other early political economists as well as the economists of Robbins's time? What could link the various things that "economists do" and might guide us as to which should be given primary stress, and which can be left for advanced study? Economics emerged along with *modern* society. Physics, architecture, and meteorology can trace their studies back for thousands of years: economics at most a few centuries. Why would a study of economics emerge at the particular time it did, and not earlier?

Among the many ways that many modern societies differ from most past societies we may mention two: affluence and interdependence. We could mention other ways as well – widespread literacy and education, the great role of science and technology – and of course they all are interrelated, but the first two are particularly important for the study of economics. Affluence is obviously of interest to economists, and as we will see in this book, interdependence is a condition that is necessary for our affluence, although it can also sometimes be a threat to it. Accordingly, we will be particularly interested in the relation of interdependence to affluence – how people work together, interdependently, and how this working together produces affluence but also poses problems we must solve.

By affluence we mean that in Europe, North America, Australasia, growing sections of Asia and Latin America and some parts of Africa, many people have access to goods and services that would have been envied by a king a few generations ago. As to the basic needs of human life – food, shelter, and clothing – they are amply provided with comfort that no one would have commanded in most earlier societies. In most of human history, obesity was rarely a problem, and never a problem among the poor. Rather, the weight problem for poor people was that they were too thin, and could not get enough food to maintain their bodies. Thomas Malthus wrote:(5) "The sons and daughters of peasants will not be found such rosy cherubs in real life as they are described to be in romances . . . the lads who drive plough, which must certainly be a healthy exercise, are very rarely seen with any appearance of calves to their legs: a circumstance which can only be attributed to a want either of proper or of sufficient nourishment." Our homes are larger than the huts in which our ancestors mostly lived, and in most cases far more comfortable than anything a king could have aspired to! In an English folk song,[(6)] a coal-miner sings that "When I first went down to the dirt [the mine], I had no cowl nor no pitshirt. Now I've gotten two or three – Walker Pit's done well by me!" *Two or three work shirts* was the beginning of affluence in an early modern coal-mining community. In addition to basic needs, many people in modern societies have access to vehicles for personal transportation, something that only the privileged would have had in earlier societies; and we enjoy a wide variety of other goods, comforts, and services. How is it that we are so fortunate?

The other characteristic of modern societies that we call attention to is interdependence. Here we may quote from Adam Smith:[(7)]

> The woollen coat, for example, which covers the day-labourer, as coarse and rough as it may appear, is the produce of the joint labour of a great multitude of workmen. The shepherd, the sorter of the wool, the wool-comber or carder, the dyer, the scribbler, the spinner, the weaver, the fuller, the dresser, with many others, must all join their different arts in order to complete even this homely production . . . what a variety of labour is requisite in order to form that very simple machine, the shears with which the shepherd clips the wool. The miner, the builder of the furnace for smelting the ore, the seller of the timber, the burner of the charcoal to be made use of in the smelting-house, the brickmaker, the brick-layer, the workmen who attend the furnace, the mill-wright, the forger, the smith, must all of them join their different arts in order to produce them . . . without the assistance and co-operation of many thousands, the very meanest person in a civilised country could not be provided, even according to what we very falsely imagine the easy and simple manner in which he is commonly accommodated.

The interdependence that Smith observed has only increased in the following two centuries.

Smith can also provide us with a key clue about the relation between these two great phenomena of modern society, interdependence and affluence. He tells us that the one is the cause of the other. He began his great book by observing that the standard of living of any country depends on the productivity of labor in that country, and that this in turn depends primarily on the "skill, dexterity, and judgment with which labour is applied." But Smith was more specific about the causes of the relatively high standards of living he observed in the most productive countries. He said it was a result of "the division of labour." Division of labor is a specific form of interdependence, and while interdependence may have other forms that are no less important, what we see in Smith's idea is true and is a central idea this book means to put forward: interdependence, in many forms, is the cause of the affluence we enjoy.

Smith's book was not the first written on topics related to economics, but it marked the emergence of economics as a discipline. We have asked why economics did not emerge until early modern times. The answer we now see is that economics is a study of phenomena that are especially prominent in modern society – affluence, interdependence, and the links between them. This book will argue that it is the job of economics to understand these things.

COOPERATION AND INTERDEPENDENCE

Some concepts from game theory will help us to understand them. "Game theory" is a misleading phrase – the subject might better be called "interactive decision theory."[8] Game theory, then, is concerned with situations in which a number of people make decisions, and the outcomes for each of them depend on the decisions made by all. This is one way to think about interdependence. Game theory can be quite a mathematical study, and its mathematical development has given rise to powerful results. For our purposes, however, broad concepts will be more important. An important distinction is between non-cooperative and cooperative decision-making. This distinction arises in non-zero sum games, that is, decision problems in which there is a potentiality for all those affected by the decision to gain more, or to lose more, depending on the decisions each may make. Cooperative decision-making is decision-making motivated by the potentiality to increase mutual gain, or to limit the mutual losses. In non-cooperative decision-making, by contrast, each person treats the decisions of others as given parameters, and is motivated by the potentiality for individual gain. If, indeed, our affluence is a consequence of our interdependence, then we have a great deal to gain by managing that interdependence well. In terms of the dichotomy of cooperative and non-cooperative game theory, that will often require that we act cooperatively.

Non-cooperative and cooperative game theory develop different concepts of decision-making, in each case assuming that only one of these two kinds of rational decision-making occurs, and assuming also that human beings are "absolutely rational decision makers whose capabilities of reasoning and memorizing are unlimited."[9] But we observe that in the real world, cooperative and non-cooperative decision-making both occur in various mixtures, and real human beings are only boundedly rational. This is why the powerful mathematical theorems of game theory will play little role in this book; but on the other hand, the opposition of cooperative and non-cooperative decision-making will be central to the book.

Why do people sometimes act cooperatively, and sometimes non-cooperatively? More particularly, why do people sometimes act non-cooperatively when they can realize mutual benefits by acting cooperatively? If indeed they were "absolutely rational decision makers whose capabilities of reasoning and memorizing are unlimited," then they would act cooperatively whenever they can benefit by doing so. To act cooperatively requires that they coordinate their decisions though, and this may be costly or impossible. The mainstream economics of the mid-twentieth century had a coherent answer to the question. One element of cooperative action was recognized: bilateral contracts of exchange. This includes the exchange of labor for money. That exchange is mutually beneficial was a key idea for mainstream economics; beyond that, however, competition, not cooperation, was supposed to determine everything else.

Apart from bilateral contracts of exchange, everything determinate in mid-century mainstream economics was determined by the interaction of large groups. In large-group interactions, it would be impossible to coordinate decisions cooperatively. Most of the things important for economics – prices, quantities produced, and similar things – would be determined by the operations of large groups, and therefore non-cooperatively. For cases in which the decisions of small groups interact to determine a result, there was no consensus. Thus, oligopoly – that is, price competition among only a small group of firms – remained an unsolved problem.

Over the years this coherence has dissolved. Two of the most influential critical papers after mid-century, those of Coase and Barro,[(10)] demanded that certain failures to make cooperative decisions should be explained, perhaps in terms of the costs of doing so. Despite these critiques, much modern economic theory makes use of non-cooperative models of decision-making, with ad hoc exceptions and special-case assumptions to explain particular empirical data or conjectures. The difficulty with this is that the whole is much less than the sum of the parts – the coherence of mid-century economics has been lost.

On the other hand, in the words of game theorist Eric Maskin,[(11)] "we live our lives in coalitions," that is, in groups that are organized to realize mutual benefit. Every business firm is such a coalition. Every club, union, cooperative, and syndicate, and many other organizations, are organized around cooperative objectives. Yet there are also non-cooperative decisions, and some of them lead to inefficiency, that is, to a failure to realize the mutual gains we might otherwise obtain. If we were "absolutely rational decision-makers whose capabilities of reasoning and memorizing are unlimited," every inefficiency would be eliminated by cooperative arrangements (as Coase and Barro point out). Instead, our economy is characterized by *imperfect cooperation*. A major objective of this book will be to understand economic activity as imperfect cooperation.

What this book proposes, then, is that economics should be defined as, and in practice is, the study of cooperative arrangements, and also of the failure to bring about cooperative arrangements – in short, a study of imperfect cooperation. In simple terms, the topic of economics is how people work together.

The first and central objective of the book is to make the case that economics is best understood in the frame of that definition. The case can only be made by examples, and, in the nature of things, many of the examples are not original – if they were original they would not support the case! To the extent that the argument as a whole is novel, perhaps these examples can derive some novelty from the context. In any case, there are also some blank spots to be filled in, and a few examples of well-established propositions in economics will be examined and rejected as simply flawed in the light of this approach.

A second objective of the book is to stress the continuities that extend from classical political economy through the neoclassical, Keynesian, and modern economics of the twenty-first century. There is no "new paradigm," but a more encompassing cognitive frame. In the same spirit, the book will borrow freely, hopefully without doctrinairism, from Austrian and other heterodox traditions, including Marxism where it is helpful, and social philosophers in the social contract tradition. Game theory of both branches will play a key role throughout.

A third objective, and I regret that it can only be third, is to make the book as approachable as possible for the reader who has limited background in economics. This is not to say that it is an easy read – that is not to be hoped for – but that some elementary topics will be explained that well-trained economists know well.

Part I, "How People Work Together," is primarily concerned with the why of it – why is interdependence a cause of affluence? Part I begins by reviewing some of the ways that our relative affluence requires us to form groups around common courses of action, that is, to work together, in Chapters 2, 4, and 5. First, Chapter 2. Production provides us with opportunities to benefit by working together. Twentieth-century production theory is not especially helpful to us here, so the book returns to Adam Smith, John Stuart Mill, and the elder Austrian School for their more substantive theories of production. Chapter 4 discusses the theory of exchange in economics. Exchange is well understood in conventional economic theory, so this chapter is relatively conventional in its content. The approach of these chapters thus is largely historical, drawing from the history of economic thought. In Chapter 5, the mutual advantages of sharing risks are explored. This requires some unavoidable digression into statistics. Part I also considers some well-known *obstacles* to working together, drawing on game theory (Chapter 3).

Part II, "Information is Not Free," addresses the mixture of cooperative and non-cooperative decision-making that we observe in the actual world of economics. It will be argued that cooperative decision-making often requires more and more costly information than non-cooperative decision-making does. Chapter 6 is a forced march through a number of well-established concepts of modern economics, from externality to organization, discussing how these phenomena can be explained by the cost of information. Chapter 7 focuses specifically on organizations as imperfectly cooperative arrangements. Chapter 8 explores, as a hypothesis, the idea that a government might be understood as a cooperative organization of the whole population. Chapter 9 revisits macroeconomics in that same context, and recognizes that the organizations that drive the macroeconomy are (however imperfectly) cooperative. Chapter 10 argues that the modern system of "democratic capitalism," with its imperfectly cooperative government, has its roots in class compromise between the employing and working classes in mature capitalism.

This book will leave more questions unanswered than answered. That is as it should be. The purpose of the book is to suggest a new frame of reference for economics, and such a new frame should raise new questions and cast a new light on old ones. If readers find some of those questions challenging, then perhaps new answers will be found.

SOURCES AND READING

(1) For some discussion of economists' definitions of economics, see Roger E. Backhouse and Steven G. Medema (2009), "Retrospectives: On the definition of economics," *Journal of Economic Perspectives*, **23**(1) (Winter), 221–33, online at http://pubs.aeaweb.org/doi/pdfplus/10.1257/jep.23.1.221, as of 1 August 2012. This is also the source of Boulding's quotation from Viner that follows. (2) John Stuart Mill (1874), *Essays on Some Unsettled Questions of Political Economy*, 2nd edition, Kitchener, ON: Batoche Books. (3) On framing decisions and knowledge, see, for example, Amos Tversky and Daniel Kahneman (1986), "Rational choice and the framing of decisions," *The Journal of Business*, **59**(4) (Oct.), S251–S278. (4) Robbins's definition is from Lionel Robbins (1952), *An Essay on the Nature and Significance of Economic Science*, 2nd edition, revised and extended, London: Macmillan, originally published in 1932.

(5) Malthus's writing on the poverty of rural workers is, of course, from Thomas Malthus (1978), *An Essay on the Principle of Population*, London: Davis, Taylor and Wilks, found online at http://socserv2.socsci.mcmaster.ca/~econ/ugcm/3ll3/malthus/popu.txt as of 1 August 2012. (6) The song is "Byker Hill," and may be found online at http://www.traditionalmusic.co.uk/song-midis/Byker_Hill.htm, as of 1 August 2012. (7) This is from Smith's great book, Adam Smith (1994), *An Inquiry into the Nature and Causes of the Wealth of Nations*, New York: The Modern Library. It may be found online at http://www.econlib.org/library/Smith/smWN.html, as of 1 August 2012.

(8) This interpretation of game theory has been pointed out by Thomas Schelling (1960), *The Strategy of Conflict*, Cambridge, MA: Harvard University Press, and is a longstanding position also of Robert Aumann. See Robert J. Aumann (2003), "Presidential address," *Games and Economic Behavior*, **45**(1), 2–14. For an introduction to game theory, I, of course, prefer my own textbook, Roger A. McCain (2014), *Game Theory: A Nontechnical Introduction to the Analysis of Strategy*, 3rd edition, Singapore and Hackensack, NJ: World Scientific. On the interpretation of cooperative and non-cooperative game theories as different understandings of rationality, see Roger A. McCain (2009), *Game Theory and Public Policy*, Cheltenham, UK and Northampton, MA, USA: Elgar, Chapter 10, section 4. (9) Reinhard Selten (1975), "Reexamination of the perfectness concept for equilibrium points in extensive games," *International Journal of Game Theory*, **4**(1), 25–55.

(10) In a famous and widely honored critique of the concept of external social costs, Coase argued that the failure of the victims and generators of external costs to arrive at a cooperative agreement requires clarification and suggested transaction costs as an explanation. See Ronald Coase (1960), "The problem of social cost," *Journal of Law and Economics*, **3**(1), 1–44. Robert J. Barro (1977), "Long-term contracting, sticky prices, and monetary policy," *Journal of Monetary Economics*, **3**(3), 305–16, criticized papers in the general disequilibrium Keynesian tradition (to which he had made important contributions) on the grounds that that they omitted to explain why employers and employees failed to come to a cooperative agreement that would eliminate inefficient unemployment. The influence of Coase's critique seems to have been spent in the 1980s, but Barro's critique reinforced the call for "microfounded" macroeconomic models, which continue to be produced in dizzying, and mutually contradictory numbers. (11) Eric Maskin (2004), "Bargaining, coalitions and externalities," Plenary Lecture, Second World Congress of the Game Theory Society, Marseille.

PART I

How people work together

2. Production: the benefit of working together

Chapter 1 argued that the affluence that many people enjoy today is a consequence of the interdependence that we also observe in modern society. But how does this happen? As Adam Smith taught us, increased productivity of labor is a necessary condition for the affluence we enjoy, so we need a theory of production that will explain how interdependence leads to increases in the productivity of labor. However, the predominant economics of the twentieth century did not provide the explanatory theory of production that we need. For that we have to go back to the economists of the 1700s and to some unorthodox economists, the Austrian School of thought. Accordingly, this chapter will focus mainly on production, and will return to the earlier schools of thought for the insights they offer.

ADAM SMITH ON WORKING TOGETHER

As we saw in Chapter 1, Adam Smith identified(1) the wealth of a nation with the productivity of labor, a consequence of the "skill, dexterity, and judgment with which labour is applied." This was a new idea at a time when most people would have identified the "wealth of a nation" with its stockpiles of gold. Smith went on to say that this "skill, dexterity, and judgment" was, in turn, a consequence of interdependence, in the specific form of division of labor. According to Smith, in the very first sentence of Chapter I, "Of the Division of Labour": "The greatest improvement in the productive powers of labour, and the greater part of the skill, dexterity, and judgment with which it is any where directed, or applied, seem to have been the effects of the division of labour." The meaning of the term "division of labour," may not be completely clear, as Smith understood, so he clarified the idea with an example that has become famous, the example of the pin factory. We can do no better than to quote it extensively:

> But in the way in which this business [manufacture of pins] is now carried on, not only the whole work is a peculiar trade, but it is divided into a number of branches, of which the greater part are likewise peculiar trades. One man draws out the wire, another straights it, a third cuts it, a fourth points it, a

> fifth grinds it at the top for receiving the head; to make the head requires two or three distinct operations; to put it on, is a peculiar business, to whiten the pins is another; it is even a trade by itself to put them into the paper; and the important business of making a pin is, in this manner, divided into about eighteen distinct operations, which, in some manufactories, are all performed by distinct hands . . . they could, when they exerted themselves, make among them about twelve pounds of pins in a day. There are in a pound upwards of four thousand pins of a middling size . . . [Moreover] Each person, therefore, . . . , might be considered as making four thousand eight hundred pins in a day. But if they had all wrought separately and independently, and without any of them having been educated to this peculiar business, they certainly could not each of them have made twenty, perhaps not one pin in a day; that is, certainly, not the two hundred and fortieth, perhaps not the four thousand eight hundredth part of what they are at present capable of performing, in consequence of a proper division and combination of their different operations.

Smith recognized the importance of machinery, which was growing in the early stages of mechanization of industry, but he also observed that it came along with the division of labor. He listed three "circumstances" through which the division of labor leads to increased productivity: (1) each person learns to be much more effective in his particular task than a generalist could be, (2) no time is lost in switching from one task to another, and (3) the simplification of tasks resulting from division of labor would make mechanization relatively easy. Thus, he saw mechanization as resulting from the division of labor, and not vice versa. "Many improvements have been made by the ingenuity of the makers of the machines, when to make them became the business of a peculiar trade [through the division of labor]; and some by that of those who are called philosophers." Important as machines might be, it was, for Smith, the division of labor into different tasks for different people, *in itself*, that led to the increases in labor productivity that, in turn, result in "that universal opulence which extends itself to the lowest ranks of the people."

We should point out something that Smith did *not* have in mind. Common sense suggests that division of labor has a fourth advantage: people can be assigned to the jobs that best suit their talents. Smith did not agree with this. Indeed, he thought it was the other way around:

> The difference of natural talents in different men is, in reality, much less than we are aware of; and the very different genius which appears to distinguish men of different professions, when grown up to maturity, is not upon many occasions so much the cause, as the effect of the division of labour. The difference between the most dissimilar characters, between a philosopher and a common street porter, for example, seems to arise not so much from nature, as from habit, custom, and education.

Smith subscribed to an idea that was called environmentalism in the eighteenth and nineteenth centuries. Since the word environmentalism has a very different meaning today, we shall call the older idea "nineteenth-century environmentalism." Nineteenth-century environmentalism held that human character is *entirely* determined by the environment in which people are reared, and not at all determined by their innate nature.

Here is the point: according to Smith it was the division of tasks *in itself* that accounted for the increased productivity of labor. As Smith also observed, the division of labor could be more easily understood in smaller enterprises, such as the pin factory, because in larger enterprises it could become more complex. As the number of tasks increases, the number of possible relationships among the tasks rises much more rapidly. All in all, the division of labor is a complex phenomenon, a phenomenon of complexity. In what follows, Smith's theory of production will be thought of as a theory of complex combination of labor.

Smith's theory remained controversial among economists in the nineteenth century. Malthus rejected it, not as untrue, but as irrelevant. Malthus (see below) conceded that the division of labor might be important in manufacturing, but not in agriculture, and he argued that the standards of living of most of the population would be limited by agriculture. Nevertheless, Smith's ideas on the division of labor were very influential in most of the nineteenth century.

In his *Principles of Political Economy*, John Stuart Mill[(2)] returned to Smith's discussion of the division of labor. Mill was probably the greatest economist-philosopher of the nineteenth century (though some would reserve that title for Karl Marx). Mill thought of his book as a summary of the discoveries of economists from Smith to the middle of the nineteenth century, but he clearly thought of the division of labor as a (if not the) fundamental cause of improving standards of living. Mill used the term "complex cooperation" to designate what Smith had called "division of labour." Mill stresses that complex cooperation is largely unconscious, and that people are unaware of it:

> In the present state of society the breeding and feeding of sheep is the occupation of one set of people, dressing the wool to prepare it for the spinner is that of another, spinning it into thread of a third, weaving the thread into broadcloth of a fourth, dyeing the cloth of a fifth, making it into a coat of a sixth, without counting the multitude of carriers, merchants, factors, and retailers, put in requisition at the successive stages of this progress. All these persons, without knowledge of one another or previous understanding, co-operate in the production of the ultimate result, a coat. But these are far from being all who co-operate in it.

In short, he quotes, in turn, Wakefield: "in order to perceive it, a complex operation of the mind is required."[(3)] Mill and Wakefield recognized that the theory they shared with Smith was a theory of "complex combination of labor."

Mill observed an implication of the division of labor that earlier writers seem to have missed: if the division of labor is important, then production will often be more efficient when it is undertaken on a large scale. If more people are employed, then labor can be divided into more distinct tasks, and the more complex division of labor can lead to higher labor productivity, so that output increases more than in proportion to the labor force. Mill wrote:

> From the importance of combination of labour, it is an obvious conclusion, that there are many cases in which production is made much more effective by being conducted on a large scale . . . The larger the enterprise, the farther the division of labour may be carried. This is one of the principal causes of large manufactories. Even when no additional subdivision of the work would follow an enlargement of the operations, there will be good economy in enlarging them to the point at which every person to whom it is convenient to assign a special occupation, will have full employment in that occupation.

In most of the twentieth century, this idea was largely relegated to the background, but it has also played an important role in much economic research in the last quarter of the twentieth century and since. In modern economics, this is called production with economies of scale or increasing returns to scale. Generally, suppose that all inputs to production are increased by the same proportion, if outputs increase in the same proportion, we say that there is constant returns to scale. If outputs increase less than in proportion with inputs, we say there is decreasing returns to scale. If outputs increase more than in proportion with inputs, we say there is increasing returns to scale; and the importance of division of labor led Mill to think that this would be the case in many branches of production. Mill also argued that this tendency toward increasing returns to scale was the reason for the increasing importance of corporations in the mid-nineteenth century. In the twentieth century, Nicholas Kaldor,[(4)] among others made the same point, and increasing returns to scale played an increasing part in economic research late in the twentieth century.

As Mill observed, economies of scale could account for the success and growing importance of corporations. This was a relatively new thing at the time Mill was writing. Mill wrote "Production on a large scale is greatly promoted by the practice of forming a large capital by the combination of many small contributions; or, in other words, by the formation of joint stock companies. The advantages of the joint stock principle are numerous and important."

Mill did not subscribe to nineteenth-century environmentalism, and is the source of the "common sense" idea mentioned above. Mill writes "The greatest advantage (next to the dexterity of the workmen) derived from the minute division of labour [is] the more economical distribution of labour, by classing the work-people according to their capacity." This does seem to be an improvement on Smith's discussion.

Mill and Wakefield had both been influenced, in turn, by the writing of George Poulett Scrope, a geologist and political economist who is hardly remembered at all now, but this section will be concluded by quoting (as Wakefield did) from Poulett Scrope's 1833 *Principles of Political Economy*:(5)

> In fact, wherever this system has made any considerable progress, the society assumes emphatically a *co-operative* character. Every member is dependent on the aid of others in everything that he does, and for everything that he enjoys. The ploughman cannot turn a furrow without the help of the wheelwright and smith; these can do nothing without that of the timber and iron merchant, the miner, and the smelter. These again must be assisted by the rope-maker, the powder manufacturer, the engineer, the carrier, and several others; while all depend upon the baker, the mealman, the butcher, the farmer, the grazier, &c. for their supplies of food; and on the tailor, the cotton and cloth weavers, the flax and wool grower, the importer of cotton, &c. for their clothing. All society is, in fact, one closely-woven web of mutual dependence, in which every individual fibre gains in strength and utility from its entwinement with the rest.

MALTHUS VERSUS THE POWER OF KNOWLEDGE

Thomas Robert Malthus(5) is not always thought of as an economist, but he was in fact the first person in England appointed to an academic position as an economist. As we have seen, he regarded Smith's theory of the division of labor as irrelevant, as it would not (Malthus felt) be applicable to agriculture. Malthus was a brilliant man – probably one of the finest minds among all economists – but he was a pessimist on principle. Malthus saw the French Revolution as a bloody disaster, and saw the disaster as being a result of a mistaken optimism that held that the conditions of ordinary people could be improved. He was determined to oppose this sort of optimism. He also saw that optimism in Smith's economic ideas, and to that extent, was critical of them.

Malthus's focus was on *agricultural* production, and it seems clear on the face that, in given circumstances, agricultural production is limited by the availability of land. It is also limited by the availability of labor, of course, and Malthus saw the *relative* availability of these two resources as the key factor in determining the productivity of labor. The more labor is applied per unit of land, the lower the productivity of labor would be. This is Malthus's law of diminishing returns. We might say that, in Malthus's view, the productivity of labor is determined by the mix of labor and other (land) inputs in the recipe, so, by contrast with Smith's "complex combination of labor" theory, we will say that Malthus had a "simple combination of different resources" theory of production.

In modern economics, the law of diminishing returns is framed in terms of marginal productivity. Briefly, we may say that the marginal productivity of labor is the addition to output if the work force is increased by one. In modern economics marginal productivity is seen as an upper limit on wages, at least so long as production is organized on the profit principle. The law of diminishing marginal productivity says that if all inputs are fixed except one variable input, then beyond some critical quantity of the variable input, the marginal productivity of the variable input will decline. The critical quantity of the variable input may be zero, in which case the marginal productivity is always decreasing, or it may be a positive number beyond which marginal productivity declines.

Malthus applied that idea to the national economy as a whole. The land available to the nation would be a constant. If the population increases, then the average output per worker declines. Thus, on average, the larger population is poorer. In particular, laborers are relatively more plentiful, so the price of their labor, their wage, declines even faster than the average product. (The marginal productivity limit on wages would reinforce this idea, if Malthus had known about it.) On the other hand, land becomes relatively scarce, so its price, the rent of land, would rise. As a consequence of rising rent, profits would decline, and so the motive for investment would disappear. As Malthus saw it, the increase in the population would lead to a stationary state, in which working people would get a wage no more than subsistence, and most of the rest of the national production would flow to landowners as rent.

But Malthus's prediction of future poverty was wrong. Over the two centuries that have passed since he wrote, populations have increased by multiples in many countries, and, on the whole, standards of living have risen. There are three ways we can interpret that evidence. First, it may be that Malthus was just wrong about the law of diminishing returns. Second, it may be that the law of diminishing returns is correct, when it is correctly applied, but that Malthus did not apply it with enough care, in that he overlooked something equally important. Third, it might be held that the last two centuries have been an exceptional period, and that Malthus will prove to be right if we wait long enough. There are some modern Malthusians who take the third view, but most economists would probably feel that two centuries are a long time to wait for the evidence to come in. Most economists subscribe to the second interpretation. What Malthus overlooked was the growth of knowledge.

Malthus was aware that new and better methods of cultivation could work against the impoverishment brought about by population growth. But he didn't think that the growth of knowledge would be a strong enough influence to prevent the increasing poverty of the population. This is where he seems to have been wrong. Malthus seems to have thought of the growth of knowledge as something of an accident that would not be continued over time. Many modern economists would argue that (at least under the right circumstances) increasing knowledge can be one of the *products* of the economic process as a whole, something that can be continued so long as the circumstances are right.

Modern economists have, on the whole, followed Malthus's approach to production theory rather than that of Smith or of other economists. Here is the reasoning. First, it takes time for technology to change, so at any particular time, in the *short run*, the technology is given. Second, so far as the allocation of resources is concerned, division of labor and similar details of the production may not matter very much. They can be left to the industrial engineers. Instead, a modern economist would define a production function. The production function is a mathematical relationship that gives us the answer to the following question: whatever the quantities of distinct kinds of inputs may be, what is the *greatest* output that can be produced, with the knowledge available at the time? With this approach we need not make any assumptions about matters of industrial engineering, only assuming that the industrial engineers do the best job they can. This upper limit determines what our production opportunities are, at a particular time, regardless of whether the upper limit is attained by division of labor or by some other quite different approach. Third, and finally, we apply the production function approach primarily to the individual firm or producing organization.

Nevertheless, this approach has some shortcomings. We have already seen one of them: in order to account for what we observe in the economic history of the last few centuries, we have to understand that the production function is constantly shifting, being moved upward by technical progress. But we cannot say anything about how the technical progress comes about. Remember, we are leaving all that to the industrial engineers! It does not seem very satisfactory to have a theory of production that ignores what seems most important about increasing production. Second, the production function approach can be mathematically quite complicated, especially if there are increasing returns to scale. Remember, if division of labor is important, then we would expect to see increasing returns to scale in many cases. But this makes the mathematics much more complicated. So it is quite common to *assume* that returns to scale do *not* increase, in order to keep the mathematics simpler. But this contradicts the whole reason for using the production function approach, which is to avoid making assumptions about industrial engineering. When we assume that returns to scale do not increase, we join Malthus in simply rejecting Smith's theory of production, whether we realize what we are doing or not.

Something else that is missing from Malthus's vision is machinery and similar produced means of production. In modern economics all products of labor and other resources that are retained for use in further production are called capital goods.* Machinery constitutes one category of capital goods, along with buildings, grapevines, cattle on the hoof, and similar productive assets. Much modern economics assumes in effect (often without ever saying it) that capital goods are a perfect substitute for land. Capital goods will play a key part in the next section of this chapter.

MENGER AND THE POWER OF COMPLEXITY

As we have seen, Adam Smith proposed what we may call a complex combination of labor theory of production, while Malthus and many modern economists subscribe to a simple combination of different resources theory. But complexity also plays a key part in the production theory associated in the late nineteenth and twentieth centuries with the Austrian School of thought in economics. This production theory originates with the great founder of the Austrian School, Carl Menger. Menger distinguished goods of the first order, which are goods that directly satisfy human wants, from goods of the second, third, and higher orders, which do not directly satisfy human wants, but can be combined to produce goods of the first order, that thus *indirectly* satisfy human wants. He wrote:[(6)]

> [I]n addition to goods that serve our needs directly (and which will, for the sake of brevity, henceforth be called "goods of first order") we find a large number of other things in our economy that cannot be put in any direct causal connection with the satisfaction of our needs, . . . next to bread and other goods capable of satisfying human needs directly, we also see quantities of flour, fuel, and salt. We find that implements and tools for the production of bread, and

* Marxists use the term "capital" differently than most other modern economists. This book will try to avoid the use of the term "capital" without qualifications that make the meaning clear, as in "capital goods."

> the skilled labor services necessary for their use, . . . [these] goods of second order [have] an indirect causal relation with the satisfaction of our needs.

Moreover, second order goods are themselves produced using other goods, such as machinery: "the grain mills, wheat, rye, and labor services applied to the production of flour, etc., appear as goods of *third* order, while the fields, the instruments and appliances necessary for their cultivation, and the specific labor services of farmers, appear as goods of *fourth* order." There is, however, little to be gained by distinguishing goods of second, third, and higher orders. It will be sufficient to distinguish goods of the first order, which directly satisfy human wants, from higher order goods, which satisfy them only indirectly. As Menger noted, machinery is a category of higher order goods, but raw materials and labor and other kinds of goods are also included in the category.

Menger criticized Smith's theory of production, not as wrong, but as incomplete. He wrote that Smith:

> cast light, in his chapter on the division of labor, on but a single cause of progress in human welfare while other, no less efficient, causes have escaped his attention . . . Assume a people which extends its attention to goods of third, fourth, and higher orders, . . . If such a people progressively directs goods of ever higher orders to the satisfaction of its needs, and especially if each step in this direction is accompanied by an appropriate division of labor, we shall doubtless observe that progress in welfare which Adam Smith was disposed to attribute exclusively to the latter factor.

That is, labor productivity is increased when production is more roundabout, relying more on higher order goods rather than on the direct production of first order goods by means of labor alone or by labor assisted only by very few higher order goods.

Ultimately, of course, goods of all orders derive from labor and natural resources such as land, as Malthus observed. However, Smith tells us that if the labor is combined in complicated ways involving division of labor, this will create opportunities for increased productivity. Menger is telling us that if labor and other resources are combined in complex ways involving a relatively great reliance on higher order goods, that will create opportunities for increased productivity. Both are complex combination of labor theories.(7)

As Menger and his successors point out, one of the characteristics of a modern economy is a vigorous trade in buying and selling higher order goods. Some of these higher order goods are durable. For this and other reasons, production that is more roundabout will generally take more time, or be continued over a longer period of time. As a result, the passage of time, in itself, becomes one of the determinants of productivity. It follows that roundabout production can pay a premium to those who are willing to wait for their payoffs in some foreseeable future time. This premium is called interest. These points were especially stressed by one of Menger's successors, Eugen von Böhm-Bawerk.[(8)]

Another detail of production by means of higher order goods is that higher order goods are complementary. They may not be perfect complements but they do complement one another to an important degree, and that in itself partly accounts for the relative productivity of roundabout production. Menger made this point quite clear, writing that if a "person has command of the flour, salt, yeast, labor services, and even all the tools and appliances necessary for the production of bread, but lacks both fuel and water . . . he no longer has the power to utilize the goods of second order in his possession for the satisfaction of his need, since bread cannot be made without fuel and water," so that the flour, salt, yeast and labor lose their character as goods. This complementarity also leads to increasing returns to scale, as another successor of Menger, Friedrich von Wieser, pointed out.[(9)] Wieser writes:

> According to Menger, . . . the farmer who loses his cart-horse loses only the value of the animal, whereas, according to our conception . . . not only does he lose the value of the animal, but he suffers, beyond this, some disturbance in the value of his remaining productive wealth . . . It is a generally-accepted fact that every productive factor furnishes the basis, not only for its own value, but also for that of all the other factors in the production.

Wieser recognized that this could create problems for the estimation of the marginal productivities of the various higher order goods, and proposed a non-mathematical solution, but the problem is really a mathematical one.

Menger and the other Austrian economists put great stress on higher order goods that are bought and sold in markets, but not all higher order goods are bought and sold. Think again of Smith's pin factory: "One man draws out the wire, another straights it, a third cuts it, a fourth points it . . . " If the pointed pieces of wire are nth order goods, then the cut but unpointed pieces of wire are $n+1$ order goods, and the straightened but not cut wire is an $n+2$ order good. In principle the one man who cut the wire might have sold the cut pieces to the man who pointed it, who could have sold the pointed pieces of wire on to the person who put the head on the pin – but that was not done in the pin factory Smith described, probably because it just would not have been convenient.

In fact, the putting out system of manufacturing was organized somewhat in that way: in manufacturing muskets, for example, a manufacturer would contract with an independent craftsman working in wood to make the stocks, sell him the materials, and buy his products; would contract with a craftsman working in brass to make the trigger assembly, sell him materials, buy his products, and so on, and after having all the parts made on contract, the manufacturer would contract with yet another craftsman to assemble muskets, selling him the parts and buying the finished muskets from him. The putting out system was common in the early days of industrialization, but was largely replaced by the factory system in which the craftsmen were hired for wages and worked under a single roof. When we think of the putting out system, though, it begins to seem that the boundary between division of labor on the one hand, and roundabout production using higher order goods on the other hand, is rather blurred. There are two ways to coordinate complex production – through organization, as in the factory system, or through markets. When complex production is coordinated through markets, the production of higher order goods is more obvious, and when it is coordinated through organization, division of labor is more obvious. The choice between these two different forms of coordination will be discussed in a later chapter.

The Austrian tradition has been continued in the twentieth century by a number of distinguished scholars(10) and still has some minority following among professional economists and a more important popular following. Of course, Austrian economists and those who share some of their ideas have tended to stress the advantages of complex roundabout production, but complexity also has its dangers. We can find one example of this in Menger's writing in *Principle of Economics*:

> When, in 1862, the American Civil War dried up Europe's most important source of cotton, thousands of other goods that were complementary to cotton lost their goods-character. I refer in particular to the labor services of English and continental cottonmill workers who then, for the greater part, became unemployed and were forced to ask public charity.

The interdependence that comes with highly efficient modern production also means that a crisis in one branch of the world economy can spread by contagion to many other regions and activities; and labor, which can be said to be the highest order good (since it is used in production of all goods of whatever order) is especially vulnerable to loss of goods-character, that is, market value, as a result. We observed this sort of contagion especially in the crisis of 2008 and the period that followed, and the Austrian production theory seems to provide a much clearer understanding of the benefits and dangers of interdependency than does the more widespread modern economics based on Malthus's theory of production.

PRODUCING THINGS AND PRODUCING KNOWLEDGE

We have seen that the growth of knowledge plays a crucial role in the improvement of human standards of living. From the Malthusian viewpoint it is this (unexplained) growth of knowledge that offsets the diminishing returns to labor in recent economic history. From the point of view of the complex combination of labor theories, new knowledge makes us aware of new ways to divide labor and new kinds and combinations of higher order goods (including labor and land), some of which lead to further increases in productivity and so in standards of living. But – Malthusian thinking perhaps to the contrary – new knowledge does not drop from the sky. New knowledge is produced by human labor and other higher order goods, work of research, reflection, experimentation, design, and development. This knowledge is itself a higher order good!

However, as we have seen, knowledge is different in some ways from other higher order goods. Consider the contrast between knowledge, as a higher order good, and the raw cotton Menger uses in his example of the impacts of the American Civil War. If one weaving shop uses a particular bale of cotton, spinning and weaving it into cotton cloth, then no other weaving shop can use the same bale of cotton. On the other hand, if one of these weaving shops uses power looms of a particular design, there is nothing to keep other weaving shops from using looms of the same design. The design of the power looms is an item of knowledge and a higher order good. In the jargon of modern economics, this knowledge is a non-rival good, while cotton is a rival good. In general, a good is a rival good if only one person can use it, and by using it deprives others of the opportunity to use it; and is non-rival if one person's use of it does not deprive others of the opportunity to use it. It is also pretty easy to prevent anyone from using the cotton if the user does not pay for it – just don't deliver the cotton if it hasn't been paid for. By contrast, it may be quite difficult to prevent people from using knowledge they haven't paid for. That is, in the jargon, cotton is an exclusive good, while knowledge is (more or less!) non-exclusive. In general, a good is exclusive if it is (relatively) easy to prevent anyone who does not pay from getting the benefit of the good, and non-exclusive if it is (relatively) difficult or impossible. Goods that are non-rival and non-exclusive are called public goods, so it is often said that knowledge of this kind is a public good.

Goods that are rival and exclusive lend themselves to being traded by buying and selling in markets. Goods that are non-rival and non-exclusive can be traded, bought, and sold only with some difficulty. This means that, other things equal, production of other goods and services is likely to be more profitable than production of knowledge; and that, in turn, leads many people (including many non-Austrian economists) to believe that it is a good idea for government policy to encourage the production of knowledge.

As we have already seen, not all higher order goods are bought and sold; some, like the straightened wire in Smith's pin factory, are used and further transformed within the business that produced them. That is sometimes the case with knowledge. When a company tries to use the knowledge it has produced, and prevent other companies from using that knowledge, the knowledge is a trade secret. The knowledge is kept a secret and, as far as possible, revealed only to people who have signed a contract not to reveal it. It can't be easy to do this, but trade secrets are quite legal and some have been kept secret for more than a century, and some very profitable products are formulated according to trade secrets. And while it is possible for trade secrets to be legitimately bought and sold, it must be very difficult and costly, and such transactions are quite uncommon.

Why would the company want to keep its knowledge a secret? The primary reason is that if the knowledge were generally known, their competitors could use it and the more informed competition would result in a decrease in the company's profits. There is a real dilemma here. On the one hand, if the knowledge were available to everybody, consumers and some other producers could benefit from that. But, on the other hand, if it were not profitable to produce knowledge, profit-seeking companies would not use their costly resources to produce knowledge. This, no doubt, is one reason why trade secrets are recognized as legitimate property. It is also the reason for the existence of patent laws.

Where patent laws exist, patenting is an alternative to trade secrecy. Since the patent is enforced by law, patents can be bought and sold and the patent can be licensed, that is, permission to use the information can be granted in return for a money payment or other consideration. This possibility can make a patent more profitable than a trade secret, especially where companies that are not direct competitors of the innovator can also make profitable use of the information. On the other hand, a patent is granted only for a limited term, and after the term is up the information is in the public domain. By contrast, a trade secret can be kept exclusive indefinitely, so long as the secret is kept. The purpose of patent laws is to make invention and innovation more profitable than they might otherwise be, and thus encourage more production of knowledge, which eventually enters the public domain for the benefit of many.

A patent law can make the production of knowledge more profitable than it otherwise would be, even if the patent law does not work perfectly. Some production of knowledge may be profitable even if there is no patent protection or trade secrecy, simply because the new knowledge can create opportunities of such importance. Trade secrets and patents may create incentives for further production of knowledge, but, given that knowledge is non-exclusive and non-rival, it is not surprising to find that much knowledge production is sponsored by government and by non-profit organizations (such as many American universities) for purposes other than earning profit. In the United States, in the twentieth century, much agricultural research was conducted by tax-supported public universities in the separate states of the American union. These same public universities have operated extension services, that is, networks of public servants in the counties assigned to bring the benefits of the university research and teaching to the farmers where they are. Of course, in the same period, a good deal of agricultural research has also been done by profit-seeking companies and protected by patents. These two sources of knowledge production seem to have been complementary, and, in any case, the system has been very successful. During the twentieth century, the rapid increase in productivity in American agriculture transformed the country, and the green revolution in other countries reflects a similar system of complementary non-profit and profit-seeking knowledge production.

Some very important knowledge production occurs as a by-product of the production of other goods and services for profit. When two goods are outputs of the same production process, such as cowhides and beef, these are called by-products. There is a great deal of evidence that labor productivity can increase simply because of learning by doing,[(11)] as experience in production increases. In the classical example, the number of labor hours to build an aircraft would decrease as the production of aircraft of the same design increased. Thus, the improvement in labor productivity was a by-product of the production of the aircraft. Of course, this learning effect is most pronounced when the methods of production are new, or, at least, new to the workers who are employed in them. But knowledge may be produced as a by-product of profitable production for sale in other ways as well. Mistakes may be made, and the mistake may turn out to be a better method than the one intended. A classic example here is that sour cream might be substituted for sweet cream in a cake recipe by accident – and the cake is found to be better for it! Moreover, experimentation and improvement of productive methods have some satisfaction in themselves, and so some knowledge production takes place just for the sake of the experience of producing knowledge. Some of the people Adam Smith called "philosophers" may be engineers working for pay, but some undoubtedly are tinkerers working for the love of the work. This human tendency should not be underestimated. Both motives – profit and subjective satisfaction – probably play a part in most knowledge production.

Thus, even in the absence of patents, trade secrecy, and government and non-profit support of knowledge production, there would be some production of new knowledge. Institutions such as patents and trade secrecy motivate some additional knowledge production above that level. Many economists and others would argue that there is unrealized potential for still more beneficial knowledge production.

Thus, ongoing production and accumulation of knowledge is a characteristic of modern economies. Adam Smith and Karl Menger notwithstanding, it may well be impossible to locate the one, fundamental cause of the improvement of productivity, as division of labor, roundabout production, or improved allocation of resources. What we can say is that modern production has a number of predictable characteristics: increasing division of labor, roundaboutness of production with reliance on higher order goods, complex interdependency, accumulation of knowledge through deliberate production of knowledge and through learning by doing and other non-deliberate improvements, market direction of production toward better allocation of resources, high levels of investment per worker, and rising labor productivity.

CHAPTER SUMMARY

The benefits of working together begin with production. In classical and Austrian economics, we find two families of theories of production: (1) complex combination of labor theories, including Adam Smith's theory that stresses the division of labor and the Austrian theory of roundabout production, and (2) the Malthusian and neoclassical theory of simple combinations of different resources. Both are correct, in that they stress different aspects that seem to be present in production in the real world. In particular, the Malthusian theory of diminishing returns yields important insights about the allocation of resources, which is an important aspect of economics. However, the complex combination of labor theories are more central to the purposes of the book, since they provide us with details to help us understand how people benefit from collaboration and cooperation. They illustrate just how we are all interdependent in a modern economy. In addition, the Austrian theory of roundabout production is helpful in that it assists us in including the production of knowledge as part of our understanding of production in general. In Austrian terms, knowledge is a higher order good; but unlike the higher order goods in traditional Austrian thinking, technological knowledge has characteristics that make it unsuitable for purchase and sale in the market, unless it is protected by patent laws.

SOURCES AND READING

The focus of this chapter is largely on ideas that we may learn from earlier generations of economics, so there is little current literature to cite. On the other hand, many of the sources are readily available online. The links listed here were all active when last checked, on 28 July 2012 or more recently. (1) The quotations from Smith in this chapter use the online edition of *The Wealth of Nations*, op. cit., Chapter 1 of this volume, note 7. This quotation is also from Chapter 1. My own choice of excerpts from *The Wealth of Nations* is available online at http://faculty.lebow.drexel.edu/mccainr/top/eco/excerpts/asmith.html. (2) An electronic edition of Mill's *Principles of Political Economy* is available at the Library of Economics and Liberty, http://www.econlib.org/library/Mill/mlP.html, as of 1 August 2012. (3) Mill was influenced by Edward Gibbon Wakefield, whose commentary appeared in an edition of Smith's *Wealth of Nations* issued in 1836. Mill's quotation is from the editor's comments on Chapter I, Book I in that edition. This commentary is accessible online at http://books.google.com/books?id=96cNAAAAQAAJ. (4) Kaldor's discussion is in Nicholas Kaldor (1934), "The equilibrium of the firm," *Economic Journal*, **44**(173) (Mar.), 60–76. The flavor of recent research assuming increasing returns to scale can be obtained from Wilfred Ethier (1982), "National and international returns to scale in the modern theory of international trade," *American Economic Review*, **72**(3) (June), 389–405. (5) Again, Scrope's book is accessible via Google Books, http://books.google.com/books?id=swVKAAAAIAAJ. Emphasis is in the original quotation. On Malthus see Chapter 1 in the current volume, note 5. (6) Karl Menger (1871), *Principles of Economics*, translated in 1976 by James Dingwall and Bert F. Hoselitz, Auburn, AL: Institute for Humane Studies. Emphasis is in the original quotation. For the quotations from Menger I rely on the archive at the Ludwig von Mises Institute. Menger's book was accessed online at http://mises.org/etexts/menger/principles.asp. (7) The synthesis of Smith's and the Austrian theories of production is not new, albeit neglected. See Richard T. Ely (1901), *An Introduction to Political Economy*, revised edition, New York: Eaton and Mains. (8) See Eugen von Böhm-Bawerk (1890), *Capital and Interest*, translated by William Smart, London: Macmillan; and Eugen von Böhm-Bawerk (1891), *The Positive Theory of Capital*, also translated by William Smart, London: Macmillan. Both are available online at the Library of Economics and Liberty at http://www.econlib.org/library/Enc/bios/BohmBawerk.html as of 22 May 2013. (9) This discussion is from Book III of Friedrich von Wieser (1889), *Der natürliche Werth* [*Natural Value*], London: Macmillan and Co, translated into English by Christian A. Malloch, edited with an introduction by William Smart in 1893, available online through the University of Toronto at http://socserv2.socsci.mcmaster.ca:80/~econ/ugcm/3ll3/wieser/natural/natural3.txt. (10) Ludwig von Mises and Friedrich Hayek were important in the twentieth century and are together known as the younger Austrian School. Hayek was honored by the Nobel Memorial Prize for Economic Sciences in 1974. The Society for the Development of Austrian Economics organizes ongoing discussion of Austrian economics: see their website at http://www.sdaeonline.org/. (11) The phenomenon of "learning by doing" in production was first observed in the production of airframes. See Harold Asher (1956), *Cost–Quantity Relationships in the Airframe Industry*, RAND Report No. 291, Santa Monica, CA. A classical discussion is from Nobel Laureate Kenneth J. Arrow (1962), "The economic implications of learning by doing," *Review of Economic Studies*, **29**(3), 155–74. This was an active field of research from the 1960s through the 1980s, but does not seem to have been so prominent in recent research.

3. Game theory: problems of working together

In Chapter 2, we have seen how the high standards of living characteristic of modern societies emerge from the complex collaboration of their members. Division of labor, roundabout production, and the production of knowledge all exemplify the kinds of interdependencies that make us prosperous. All this complexity and interdependence also creates opportunities to "game the system," and for that reason we now turn to the study of game theory. As we noted in Chapter 1, the term game theory is not the best description of the field. Both Robert Aumann and Thomas Schelling have said at different times that what we call "game theory" would be better described as "interactive decision theory."(1) In our "closely woven web of mutual dependence,"(2) each of us makes decisions that affect the activities and well-being of others, so interactive decision theory is crucial for our understanding of economics, and plays a great role in recent economic ideas and research.

Game theory has two major subdivisions: non-cooperative and cooperative game theory. In cooperative game theory, we assume that people can always make and carry out agreements to choose a joint course of action, either to obtain mutual benefit from their joint action or to "gang up" on other groups. In non-cooperative game theory we assume that people can make and carry out agreements of that kind only if there is some enforcement of the agreement, such as a legally enforceable contract.

Game theory was founded in a pathbreaking book by John von Neumann and Oskar Morgenstern.(3) Von Neumann was one of the foremost mathematicians of the twentieth century and Morgenstern an economist educated in the Austrian tradition. Their approach was cooperative and they began the cooperative tradition in game theory. However, many economists believe that non-cooperative game theory is more realistic. In any case, cooperative game theory is more complicated and difficult, so applications of non-cooperative game theory are better understood. Accordingly, we will begin with non-cooperative game theory, and explore the cooperative approach in later sections of the chapter. The most famous example in non-cooperative game theory is one in which self-interest leads to a frustrating result. It is known as the Prisoner's Dilemma.

HOW SELF-INTEREST DENIES ITSELF

John von Neumann was at the Institute for Advanced Study in Princeton, New Jersey and Morgenstern was on the faculty of Princeton University when they wrote their book, *Theory of Games and Economic Behavior.* Other Princeton mathematicians became interested, including Albert Tucker and some of his graduate students. Tucker wanted to explain this interesting new field to an audience of non-mathematicians and originated the Prisoner's Dilemma example.[(4)]

The example begins when two suspicious characters are captured near the scene of a burglary. They are taken to separate interrogation rooms where they cannot communicate with one another, and questioned separately. Each is told that if he confesses and implicates the other, and the other man does not confess, then the one who confesses will be released. The other will go to prison for ten years. If neither confesses, they can be convicted of a lesser crime, and both will serve two years. If both confess they will both serve six years. Their decisions and the results are summarized in Table 3.1. In the table, the First Burglar's decision to confess or refuse to confess determines whether the outcome will be in the third row or the last row, and the Second Burglar's decision to confess or to refuse determines whether the outcome will be in the third column or the last column. The outcome of their interactive decisions is described by the two numbers shown in the cell in the row and column corresponding to their decisions. The first number is (minus) the number of years served by the First Burglar and the second number is the negative of the number of years served by the Second Burglar. The numbers are shown as negative numbers because they are penalties, not rewards.

Table 3.1 The Prisoner's Dilemma

First Penalty to First Burglar, Second Penalty to Second Burglar		Second Burglar	
		Confess	Refuse
First Burglar	Confess	–6,–6	0,–10
	Refuse	–10,0	–2,–2

The "dilemma" for a prisoner in this example is whether to confess or not. Let us try to imagine how he might think it through. "One of two things will happen: either the other prisoner will confess or he will not. Suppose he does confess. Then I am better off to confess, for six years in prison, than to refuse for ten years in prison. In that case, to confess is my *best response* to the other prisoner's decision to confess. On the other hand, suppose he does not confess. Then I am better off to confess, implicate him and go free than to refuse to confess and go to prison for two years. Once again, confession is my best response to the other prisoner's decision. Therefore, I will confess." Both prisoners reason in this way, and as a result they both confess and go to prison for six years.

This example illustrates several things about non-cooperative game theory:

- The outcome of the decisions of all of the "players in the game" is a list of payoffs, with one payoff for each agent in the game. In this case the "payoffs" are negative since a period in prison is a negative "payoff."
- We assume that the agents are rational in that they each choose the best response to the decision made by the other or, if there are more than two, to the decisions made by the others.
- The first two points make sense only if the "payoffs" are at least roughly proportionate to the subjective states of mind that motivate the decisions, that is, to the utility the agent derives from the outcome of the interactive decision. Accordingly, von Neumann and Morgenstern proposed a method of measurement of utility, and many game theorists rely on it.
- It is a non-constant sum game. That is, the sum of the payoffs to both prisoners depends on the decisions the two prisoners make, and can vary from -4 to -12. Von Neumann's earlier work on game theory had assumed that the payoffs must add up to zero, as they do in "friendly" gambling games. When the payoffs add up to a constant number such as zero, we would describe the game as a constant sum game. Otherwise, as with the Prisoner's Dilemma, it is a non-constant sum game. Non-constant sum games may have win–win or (as in this case) lose–lose outcomes. This is particularly important for economics, as we have seen in the previous chapter.
- The alternatives among which the interacting agents must choose are often called strategies. In this case we see that there is one strategy, to confess, that is the best response to any strategy that the other agent might choose. A strategy with that property is called a dominant strategy.
- If the two prisoners had been less "rational" they might have done better. Suppose, for example, that they had both irrationally believed that some supernatural power would punish them if they were to confess. Acting on this irrational belief could leave them with two years in prison instead of six.

It is this last point that has attracted much of the attention to the Prisoner's Dilemma. This example shows us how even a fairly simple kind of interactive decision can frustrate the intentions of the decision-makers. A dominant strategy is a rational, self-interested strategy in an obvious sense that is difficult to dispute. Nevertheless, when both prisoners act in this rational, self-interested way, the result is something that no rational, self-interested burglar would choose if he could avoid it. We should observe that a dominant strategy will not always create a dilemma of this sort – if the numbers were different the dominant strategies might lead to the best outcome for everybody. Nevertheless, in the far more complex interdependencies of a modern economy, we naturally expect that equally challenging and even more challenging dilemmas of rational decision-making will arise.

However, human beings have been living with Prisoner's Dilemmas and similar interactions for a long time. We have evolved laws and institutions to deal with them. To illustrate what this means, consider the following "Taking Game." Two protohumans each have possessions that can assure each of them utility of 5, if the possessions are not interfered with. But each one has the option of attempting to steal the other's possessions. Now, stealing is work – for the example, we suppose that the effort of stealing the other's possessions is equivalent to a deduction of 2 units of utility. Thus, if the First Protohuman succeeds in stealing the possessions of the Second Protohuman, the First Protohuman's payoff (utility) is $5 + 5 - 2 = 8$. The strategies and payoffs are shown in Table 3.2.

Table 3.2 A Taking Game

First Payoff to First Protohuman, Second Payoff to Second Protohuman		Second Protohuman	
		Steal	Don't
First Protohuman	Steal	3,3	8,0
	Don't	0,8	5,5

This game is very much like the Prisoner's Dilemma. To steal is a dominant strategy, but both will be better off if neither of them steals. In fact, we have laws and customs that enforce property rights and discourage stealing. It is not only that the law prohibits stealing – in addition, our customs and habits also discourage us from stealing. These laws, customs, and habits do not work perfectly – some theft does take place – but property is usually secure enough that we can go about our business without much worry that our property will be stolen. But the Taking Game reminds us of why our societies have evolved laws protecting property rights and punishing theft. Some scholars think that our brains have evolved in ways that help us to deal with dilemmas like the Prisoner's Dilemma and the Taking Game.

The Prisoner's Dilemma game is special in another way. Not all games have dominant strategies, as will be seen in the next section.

INTERPERSONAL KNOWLEDGE AND BEST RESPONSES

If a game has no dominant strategies, it may be even more difficult to figure out what the "rational" decisions would be. Nevertheless, the dilemmas in the previous section give us an important hint. When each of the decision-makers chooses his or her dominant strategy, each is choosing his or her best response to the strategy the other decision-maker has chosen. This can occur even if the game has no dominant strategies.

This will be clearer with an example, so we will consider a hypothetical example of knowledge production. Two companies have done preliminary research applicable to a new office communication system. We will call them Firm 1 and Firm 2. The two firms will be producing higher order goods that must be combined as components to produce a single final good. For convenience, we will call the two components the "framework" and the "extension," though the specific details of the two components are not important for the example and will not be discussed.

Communication systems often depend on compatible standards, and there are two standards that might be used: a well-established but somewhat less effective standard, and an advanced standard that is potentially more effective but less well established, so that the research to apply the advanced standard will be more costly. Each firm is planning further research to produce components for the new system, and has to decide whether their component will be compatible only with the established standard or with both the established and advanced standards. On the other hand, if either company makes its component compatible with the advanced standard as well, that company will face higher research costs. If the product is to be at the advanced standard, both components must be produced to that advanced standard. The strategies and payoffs are shown in Table 3.3.

A first point we observe is that the Knowledge Production Game does not have any dominant strategies. Suppose that Firm 2 chooses the established standard. Then Firm 1's best response to this decision is also to choose the established standard, for a payoff of 5, rather than doing the additional work to make their component consistent with the advanced standard, for a payoff of 3. But suppose that Firm 2 chooses the advanced standard. Then Firm 1's best response is to choose the advanced standard for 10 rather than 5. Since Firm 1's best response depends on the decision made by Firm 2, neither strategy is dominant. Firm 1 cannot make a "rational" decision without some information or at least a judgment or conjecture as to the decision that Firm 2 will make. This added layer of complexity often arises in non-cooperative games.

Table 3.3 A Knowledge Production Game

First Payoff Firm 1, Second Payoff to Firm 2		Firm 2	
		Advanced	Established
Firm 1	Advanced	10,10	3,5
	Established	5,3	5,5

Again, if Firm 2 chooses the established standard, Firm 1's best response is also to choose the established standard. Similarly, if Firm 1 chooses the established standard, then Firm 2's best response is to choose the established standard. When each of them chooses the established standard, each is choosing her or his best response to the decision made by the other. This situation – where each one is choosing the best response to the decision made by the other – is called a Nash equilibrium, after John Forbes Nash, whose research established the Nash equilibrium as a central concept of game theory.(5) In the Prisoner's Dilemma and the Taking Game, the dominant strategies also define Nash equilbria for those games.

If we think of the non-cooperative game as a mathematical problem, the Nash equilibrium is a definition of a solution, and is often spoken of as a solution to the game. Ideally, for each game we would like to have just one solution. That is true for the dilemmas in the previous section, where the dominant strategy solution is the only Nash equilibrium. But this is not always the case with the Nash equilibrium. The Knowledge Production Game is a case in point. We have already seen that if Firm 2 chooses the advanced standard, Firm 1's best response is also to choose the advanced standard. Conversely, if Firm 1 chooses the advanced standard, Firm 2's best response is to choose the advanced standard. Thus, a situation in which both firms choose the advanced standard is *another* Nash equilibrium for the game.

This poses an even more difficult decision problem for the two firms.(6) If the Nash equilibrium of the game is unique, as it is when there are dominant strategies (and in some other games), then each decision-maker can assume that the other decision-maker knows this, and have confidence that (supposing both are rational in a non-cooperative sense) each will choose the Nash equilibrium. In this case that kind of simplification cannot be done. The simplest way to solve the problem would be for the two firms to sit down together and share their plans, but if they were to do so they would reveal the information they have already produced in their preliminary research, and as we have seen in the previous chapter, it may be necessary to keep that information secret.* Thus, each decision-maker will have to base his or her decision on a conjecture about what the other person's decision will be – and they both know this, so Firm 1 can rationalize a decision on the basis of a conjecture about what Firm 2 conjectures that Firm 1 will do, and so on. Suppose, for example, that Firm 1 reasons "Firm 2 thinks that we are a very conservative firm, and expects that we will choose the established standard, so they will choose the established standard. Thus, we will also be better off to choose the established standard." This is an example of a rationalizable strategy.(7) (This is a technical term!) In general, a strategy is said to be rationalizable for Agent K if it proceeds from (1) an assumption about other decision-makers' conjectures about the strategy Agent K will choose, (2) the inference that they will choose their best responses to the strategy they guess Agent K will choose, so that (3) Agent K chooses his or her own best response to those strategy decisions by the other agents. In the example directly above, Firm 1 is "Agent K." Here is another possibility: Firm 2 thinks: "Firm 1 knows that we are a very dynamic firm so that we will choose the advanced system. Accordingly, their best response will be to choose the advanced system, and we in turn will be best off by choosing the advanced system." Thus, each firm chooses a rationalizable strategy – Firm 1 chooses the established standard, for the expected payoff of 5, and Firm 2 the advanced system for a disappointing payoff of 3. They discover that their assumptions have been wrong, but by then it is too late to correct their errors.

Clearly, rationalization is a pretty fallible process. But it does not always fail. In particular, every Nash equilibrium is rationalizable. Suppose Firm 1 reasons: "Firm 2 is a very progressive firm and they know we know that, so they will expect us to choose the advanced standard. Accordingly they will feel comfortable choosing the advanced standard, and our best response is to choose it too." That rationalization, taken together with Firm 2's rationalization in the previous paragraph, will lead to the advantageous Nash equilibrium with payoffs of 10,10. More than

* To be more complete, this could be treated as a two-stage game, in which the first stage is the decision whether or not to reveal the information from the preliminary research. The example shows that there may be advantages as well as disadvantages from doing so. For this particular example, we simply assume that secrecy is the best policy at that first stage.

that: the evidence will tell each of the two firms that their beliefs were correct.

But that is not the only possibility. Suppose instead that Firm 1 conjectures as in the paragraph before last that Firm 2 will respond to Firm 1's reputation as a conservative firm, and at the same time Firm 2 reasons: "Firm 1 is conservative, and they know we know they are, and thus they will expect us to adapt to that by choosing the established system. Accordingly, they will choose the established system, and we had best do the same." These conjectures will lead to the less advantageous Nash equilibrium with payoffs of 5,5. Nevertheless, once again, the two firms will have evidence that their judgments were in fact correct.

Reasoning like "Firm 2 thinks that we are a very conservative firm, and expects that we will choose the established standard" expresses a belief system for the different decision-makers. The key conclusion we want to draw is that the belief systems of the decision-makers will be confirmed by the evidence of experience whenever their decisions correspond to a Nash equilibrium. If there is more than one Nash equilibrium in the game, there will be more than one belief system that could be confirmed by experience. In the Knowledge Production Game, one belief system leads to the Nash equilibrium with payoffs at 10,10, but another belief system, which agrees equally well with the facts of experience, leads to the equilibrium with payoffs 5,5.

We should stress that *not all* of our beliefs depend on the choice of Nash equilibria! Our beliefs about the value of the gravitational constant, for example, do not seem to depend on the conjectures that other decision-makers make about us. Thus, we can rely on scientific experience to give us the unique value of the gravitational constant to a good approximation. Our beliefs *about each other*, however, are not a subject of scientific confirmation in quite that simple way. Our beliefs about one another may be confirmed by evidence and experience, not because the beliefs are simple facts, but because the beliefs lead us to act in ways that correspond to a Nash equilibrium. And that may trap us in an inferior Nash equilibrium like the one in the example with payoffs of 5,5. This is one of the enigmas that arise from our interdependency in a modern economic society, especially from the dependence of each of us on decisions made by many others.

GETTING OUR ACT TOGETHER

In all of the three "games" we have considered so far, there are Nash equilibria in which both agents have reasons to regret the decisions they have made. In the Prisoner's Dilemma, for example, it seems pretty obvious that both prisoners are better off with "Refuse, Refuse" than they are with the dominant strategies "Confess, Confess." Rather than rely on "it seems pretty obvious" though, it is time for a little systematic thinking about how we can say that one situation is better than another, when the payoffs, utilities or preferences of two or more agents vary from one situation to another.

The key idea comes to us from the Italian economist and sociologist Vilfredo Pareto.[8] Pareto was a critic of the utilitarian idea that ethical decisions and public policy should be chosen in such a way as to bring about the greatest total utility. Pareto argued that utilities just could not be added up: that being purely subjective, the utilities of different people could not even be compared. (Many economists of the twentieth century and of today agree with Pareto.) According to Pareto, even if we cannot add the utilities of the different people in the economy, we may sometimes be certain that one situation is better than another. Suppose that in situation A some person is better off (in terms of his or her own non-comparable subjective utility) than he or she is in situation B, and no one is worse off, then situation A is clearly better than situation B. We express this by saying that situation A is Pareto-preferable to situation B. Returning to the games in Tables 3.1–3.3, we can say that "Refuse, Refuse" is Pareto-preferable to "Confess, Confess" in the Prisoner's Dilemma, that "Don't, Don't" is Pareto-preferable to "Steal, Steal" in the Taking Game, and "Advanced, Advanced" is Pareto-preferable to "Established, Established," or any other outcome, in the Knowledge Production Game.

In economics we apply this concept especially to the efficient allocation of resources. If we can shift the allocation of resources in such a way that the new allocation is Pareto-preferable to the old one, that is an increase in efficiency. Suppose, then, that every potential opportunity to improve efficiency in that way has already been realized that there is no unrealized potentiality to shift the allocation of resources in such a way that the new allocation is Pareto-preferable to the old. That is, in other words, there is no way to shift the allocation of resources so someone is made better off and no one is made worse off; that in fact no one can be made better off without making someone else worse off. Then we say that the allocation is Pareto-optimal and that is our common criterion of economic efficiency.

To return to game theory, the very fact that (for example) "Advanced, Advanced" is Pareto-preferable to "Established, Established" suggests a solution to the problem. The solution is that the two firms might agree with one another to choose the advanced standard. Since "Advanced, Advanced" is a Nash equilibrium, self-interested rationality will assure that each firm acts according to the agreement. This would be a little more difficult with a game like the Prisoner's Dilemma, though. Certainly "Refuse, Refuse" is Pareto-preferable to "Confess, Confess," so the two prisoners can agree readily enough that "Refuse, Refuse" should be their joint strategy – but will they carry out the agreement? Even if the First Burglar knows that the Second Burglar can be trusted to carry out the agreement, the First Burglar's best response will be to confess – to defect from the agreement, to use the language of game theory – and both burglars know that. Defecting from an agreement one has made, in the hope of exploiting a commitment made by another, is sometimes called opportunism. Unless the two burglars trust one another and are both trustworthy, or there is some enforcement of the agreement, the agreement will not be carried out. Trustworthy behavior is not "rational" as the term "rational" is used in non-cooperative game theory, so we can narrow it down a little: if both are "rational" in this non-cooperative sense and there is no enforcement, the agreement will not be carried out.

Notice the contrast between this and the Knowledge Production Game. If the two firms agree to "Advanced, Advanced" in the Knowledge Production Game, there is no need for enforcement, because a Nash equilibrium is self-enforcing. In the Taking Game, "Don't, Don't" is Pareto-preferable to "Steal, Steal" and we have a general agreement in society, what some philosophers call a social contract, not to steal. This agreement is enforced by law, punishment, social convention, and widely accepted moral rules.

Thus far in Chapter 3, we have used the word "rational" somewhat cautiously, putting it in quotation marks or qualifying it with terms like "rational in a non-cooperative sense." In most modern economics and much game theory, "rationality" simply means what we have been calling "rationality in a non-cooperative sense." To a non-economist, it may seem a rather strange concept of rationality, since, as we have seen, two people may not be able "rationally" to carry out an agreement that makes them both better off. Put otherwise, there is no commitment in modern economic theory and non-cooperative game theory. We might suggest instead that the failure to make commitments that make one better off is one instance of a failure to act rationally. We could then concede that failures to act rationally are fairly common in real human behavior, and that includes failures to make profitable commitments among many other instances. The second part of this book will explore some important reasons why these failures to decide and commit rationally are as common as they are, and the implications of these reasons.

On this score, von Neumann and Morgenstern were pretty clearly on the opposite side from modern economics: their assumption was that if people can benefit from carrying out an agreement, rational people will find a way to carry it out. This is not an especially "nice" conception of human nature, by the way – and that shows up especially with respect to threat behavior. People may threaten to do things that would make them worse off if they carry out the threat – "This hurt me more than it hurts you." Nash, and scholars who follow his lead, argue that such threats will not be carried out, so are irrelevant and can be ignored. But if people can commit themselves to keep their agreements, they can also commit themselves to carry out threats even when the threats are costly to the threatener. Von Neumann and Morgenstern assumed that rational people would always be able to make binding commitments when it was to their benefit to do so. Thus, they would make agreements to carry out joint strategies – von Neumann and Morgenstern called these agreements coalitions – and would make the most extreme threats to increase their bargaining power as they did so. Cooperative game theory has grown out of their work and explores the implications of rationality in this different sense.

Both concepts of rationality are too extreme. People do sometimes make and carry out commitments; people do sometimes fail to carry out their commitments, both where agreements and threats are concerned; people sometimes opportunistically exploit the commitments of others; and, recognizing this possibility, people sometimes refuse to make commitments and choose instead to act non-cooperatively. On the other side, both branches of game theory have important things to teach us. We can often observe non-cooperative behavior in the world around us, but it is no less true that we observe many coalitions of people working together with common strategies in their mutual interest.

Accordingly, we will next introduce some common ideas of cooperative game theory.

VALUE CREATION

As Adam Smith taught us in Chapter 2, we can often increase our productivity by means of division of labor. Accordingly, we will begin with a game theory example suggested by his discussion of the pin factory. To keep it as simple as possible – but no simpler! – we will suppose that there are just two workers, simplifying the example that Smith had already simplified very much. Each worker has to choose among three strategies. On the one hand, each can work alone, without division of labor. If they do that, each can produce one unit of output. (This is our definition of the unit of output for the purposes of this example.) This is one strategy. On the other hand, they can work together, with division of labor. This requires that they perform different tasks, tasks that are highly complementary in ways that enhance their productivity, so that the total output from working together with division of labor can be as much as 6, with the same efforts they would make working alone. But in order to get any output at all, they have to choose different tasks. These are their second and third strategies: choose Task 1 or Task 2. If both tasks are done, then each worker receives a payoff of 3 (Table 3.4).

Table 3.4 A Division of Labor Game

First Payoff to Worker 1, Second Payoff to Worker 2		Worker 2		
		Work alone	Task 1	Task 2
Worker 1	Work alone	1,1	1,0	1,0
	Task 1	0,1	0,0	3,3
	Task 2	0,1	3,3	0,0

This game has three Nash equilibria, "Work alone, Work alone"; "Task 1, Task 2"; and "Task 2, Task 1." The four other outcomes are inferior to *all* Nash equilibria, and both of the latter two Nash equilibria are Pareto-preferable to the Nash equilibrium at "Work alone, Work alone." But, from a non-cooperative point of view, this game presents a problem of coordination. Every strategy in this game is rationalizable. However, if the two workers rely only on rationalization, some very bad outcomes are possible, and each worker can avoid this risk by working alone.

Some additional information is necessary to solve this coordination problem. But that is not very difficult: for example, the two could get together and flip a coin. Heads, Worker 1 takes Task 1 and Worker 2 takes Task 2, and tails, conversely. Another solution is to rely on a trusted third party to direct the two workers as to which task to do. Since neither of them gains by taking one task rather than the other, there is no reason for them not to follow the directions of the third party. In a more realistic example, with many workers and many tasks to be done, the trusted third party becomes a more natural solution. This role – coordinating the division of labor – is the role of the entrepreneur, according to American economist John Bates Clark.(9) But what will motivate the coordinator? We will return to this, but first let us make the example *a little* more realistic.

In Table 3.4, both workers get equal payoffs. Let us change the example in just one way: if both tasks are done, the person who does Task 2 ends up with the product in his or her possession, for a payoff of 6, the entire value of the joint productive effort, while the person who does Task 1 is left with nothing in his or her possession. Task 1 is an intermediate stage of processing or a higher order good that does not, in itself, yield marketable value, but without which the higher productivity of producing with division of labor cannot occur. These assumptions are summarized in Table 3.5.

Table 3.5 Another Division of Labor Game

First Payoff to Worker 1, Second Payoff to Worker 2		Worker 2		
		Work alone	Task 1	Task 2
Worker 1	Work alone	1,1	1,0	1,0
	Task 1	0,1	0,0	0,6
	Task 2	0,1	6,0	0,0

We see that this game has only one Nash equilibrium, and that is where both workers work alone. Moreover, there is no other outcome for this game that is Pareto-preferable to working alone. All the same, the division of labor outcomes, "Task 1, Task 2" and "Task 2, Task 1" seem clearly superior in the sense that the total payoff is tripled. But the worker who does Task 1 is worse off as a result. Suppose, then, that Worker 1 approaches Worker 2 with this proposition: "You take Task 1, I'll take Task 2, and I will pay you half, that is, 3, after the product is complete and is sold." The payment of 3 is called a side payment in game theory. The side payment will make it worthwhile for Worker 2 to take Task 1. This agreement – to choose a common strategy and to share the benefits of that strategy – is called a coalition in cooperative game theory.

There are two complications to deal with before we go on. When we include the agreement between the two workers and a side payment in the game, we have really transformed the game shown in Table 3.5 into a different, somewhat more complicated game. To make a side payment is a new strategy[(10)] that did not exist in the original game. Making a side payment can only make the payer worse off, so side payments would never be made (without enforcement) in a Nash equilibrium or in a noncooperative game, but we are now assuming that the workers can and do make binding commitments, so the offer of a side payment can assure the worker who takes Task 1 that he will not be left without anything. (A legally enforceable contract is one way that Worker 1 could reassure Worker 2 that the side payment will be made.)

There is another complication, and although it is a minor one in this context, it is more important in some less simple (more realistic) applications. We understand that people are motivated by subjective benefits

– utility or preferences – not by quantities of output or by the money the output could be sold for. In some cases, money payoffs may not correspond well enough to the subjective benefits to persuade an agent in the position that Worker 2 is in that the commitment is worth his or her while. If, for example, Task 1 were to require a much greater effort than working alone, a payoff of 3 might not be enough to compensate for that effort. But we have no reason to think that that is always the case, and here we are considering an example in which we assume that there is no difference in effort or in other subjective benefits other than the benefits that come from a higher income from selling the product. Money payoffs and subjective benefits will correspond closely enough so that an increase in money payments will always correspond to an increase in subjective benefits. If so, it is almost as if the side payment were made in units of utility. This sort of example is known in game theory as a transferable utility (TU) game. The TU assumption is, of course, a simplifying assumption and only approximately true in any case, but it seems a good enough approximation in practice in many cases where the benefit of working together is big enough to overshadow the difference between subjective and money benefits.

In a TU game, the total payoffs generated by the coalition can be distributed among the participants in the coalition in whatever way may be convenient. The total payoffs constitute the *value* of the coalition. In Tables 3.4 and 3.5, the value of a coalition between the two workers is 6. The payoffs they could obtain without a coalition total to 2, so the formation of the coalition has created value in an amount four times as great as the value of a worker working alone. We can identify value creation with the formation of coalitions to work together in production, in many cases.

There is still a coordination problem to be solved – which worker does Task 1 and which does Task 2? Once again, in a realistic game with many workers and many tasks, the most natural way to solve the coordination problem is to have the decision made by a trusted third party. To motivate the trusted third party, the workers might offer him or her a side payment – a salary – at the cost of some reduction of their own pay. The group of workers would then have organized a workers' cooperative.(11) The coordinator is then a member of the coalition, that is, the workers' cooperative. (It might be smart to base the coordinator's salary partly on his or her success in coordinating production, that is, on profits, but experience would tell whether that is necessary or not.)

Another possibility is that the coordinator might take the initiative to bring the coalition into existence, offering side payments (wages) to the workers on the condition that each of them takes one of the tasks as directed. The coordinator would keep whatever is left over, the profits, for him- or herself. In this case, the coordinator would be an entrepreneur not only in John Bates Clark's sense, but also in the sense we find in the ideas of Carl Menger. If the workers' only alternative is to work alone, then they would be better off with any wage greater than 1 and most of the value created by the coalition would go to the entrepreneur. (In practice, the coalition will not exist in isolation, and the entrepreneur who founds this coalition is likely to face competition for labor from other entrepreneurs, so that he may have to offer a wage greater than 1.)

There is yet another point that we have left in the background. In modern production, division of labor is associated with extensive use of higher order goods, that is, capital goods and raw and semi-finished materials. If an entrepreneur takes the initiative to form a coalition for production, he or she will also have to assemble the higher order goods that complement the labor committed to the enterprise. The cost of these higher order goods – including the cost of diverting consumption from the present to the future, that is, the interest that could be earned on the money put up for the purpose – will be deducted from his or her profits.

If the coalition were organized instead as a workers' cooperative, then the workers who form the coalition will also have to find money to assemble the higher order goods needed. Historically, this has been an obstacle to the formation and sustenance of worker cooperatives, and this obstacle probably explains why they have been rare. However, the problem can be solved, and there have been some important successes in this kind of organization. One of the most widely studied is a complex of cooperative organizations at Mondragon, Spain.[(12)]

What we conventionally call capitalism is an economic system in which coalitions for production are predominantly formed by entrepreneurs, and the share of employees in the value created by these coalitions is influenced primarily by competition among entrepreneurs for a limited supply of labor. In this section we have focused on the formation of coalitions for production. These will be particularly important for the purposes of this book. However, coalitions may be formed for a wide variety of purposes: to facilitate exchange, to organize recreation or worship, for mutual self-defense or attack or political campaigning or to enhance bargaining power, to conserve or share resources such as a home, a residential drive or a stream or lake, to educate the young, to make music, and so on. With a few appropriate assumptions, the abstract principles of cooperative game theory can be applied to all of these.

DIVIDING THE SPOILS

We have adopted the view that value is created by the formation of coalitions, that is, groupings of people for purposes of working together or for exchange. To keep things simple, we have assumed that the value from their collaboration can be divided among them in any way that is convenient, a simplifying assumption that is often called transferable utility (TU). How, then, will a group of rational agents in a coalition divide among themselves the value created by their collaboration? The unpleasant truth is that, in general, we do not know. There have been several attempts to answer this question in general, but there is no real consensus among game theorists as to which is the best answer. Economists tend to focus on the competitive alternatives, and at least in some cases, competition does seem to give the answer. But in other cases (cases that seem important to some game theorists) competition does not give a definite answer, although it does set limits. We will explore the game-theoretic understanding of competitive alternatives, which is expressed in game theory as the theory of the core of the game.

To illustrate the core, we need a somewhat larger example. Let us say that there are four workers. The technology behind Table 3.5 is still available to them: that is, any coalition of two workers can produce a value of 6, while an individual worker working alone can produce only 1. In addition, we suppose that the technology is such that a three-person coalition, with a slightly more complex division of labor, can produce a value of 12 and that a four-person coalition can produce a value of 18. We will not go into much detail about the strategies and payoffs for three- and four-person coalitions – for a transferable utility game, for present purposes, all that really matters is the value that a coalition can create – not the details of how the coalition creates that value. The assumptions are summarized in Table 3.6.

Table 3.6 A Four-person Cooperative Game

Coalition of Size	Value
1	1
2	6
3	12
4	18

Suppose, then, that the coalition of all four workers is formed. The average value per worker is 4.5, which is more than any smaller coalition can produce on the average. Thus, we might expect that rational workers would form that coalition. But how will they distribute the 18 units of value? Suppose, for example, that Worker 1 keeps 9 units and the rest are distributed 3,3,3. (Perhaps the formation of the coalition was his idea so he feels he can keep half.) But this arrangement will not be stable, since the other three workers can drop out of the four-person coalition, form a three-person coalition, and produce a value of 12. This is more than the payoff of 3 each – if they divide equally, they can have 4 each. Being rational, they will do that. But, being rational, and knowing that the others also are rational, Worker 1 will anticipate their secession and not demand so much. The four-person coalition will be unstable unless any three-person group makes at least as much as they can if they form a separate coalition. This can be so only if each member receives at least 4.

The idea behind the core is that a coalition and its payoffs will not be stable unless any individual and any subgroup receives at least as much as they could receive in other coalitions they might participate in, including coalitions formed by subgroups that break away from the larger group as in this case. In that sense, the core captures the idea that competitive alternatives put a lower limit under the payoff that goes to each member of the coalition. In this example, it is the possibility of forming a three-person breakaway coalition that establishes the lower limit.

The example illustrates one other important possibility. In this game, each person has to receive at least 4 in order for the game to be stable. But the coalition generates a value of 18, which is more than enough to pay 4 to each member of the coalition. Thus, there may be many different stable payoff schedules. Any payoff schedule that gives at least 4 to each member will be a payoff schedule in the core of the game and so it will be stable under competitive pressure. The remainder of 18 – 16 = 2 units of value can be distributed among the four players in any way whatever. In particular, payoff schedules that distribute payoffs as 6,4,4,4 or 5,5,4,4 are all stable and are in the core of the game, since equal distribution is not required in order for a payoff schedule to be within the core.

For the game in Table 3.6, then, there are many payoff schedules in the core. That is not always true. Suppose we change the values of the coalitions from Table 3.6 in just one small way – for the game in Table 3.7, a two-person coalition can produce 9. That would mean that any subgroup of two persons must receive at least 9, since otherwise they would drop out and form a separate two-person coalition. In order for each two-person group to receive 9, it is necessary for each individual to receive 4.5. Now, four times 4.5 is exactly 18 units of value – and the coalition of all four agents can produce no more. Thus, a payoff schedule of 4.5 units of value to each member of the coalition is the unique, stable payoff schedule for this game. In this game, the core has exactly one payoff schedule – just what we want our economic theory to give us.

Table 3.7 Another Four-person Cooperative Game

Coalition of Size	Value
1	1
2	9
3	12
4	18

But we face yet another problem. In Table 3.8, we make two slightly different changes in the game from Tables 3.6 and 3.7. In Table 3.8, we assume that a two-person coalition can produce just 8, in place of 6 or 9 in the earlier tables; and a three-person coalition can produce a value of 15, in place of 12 in Tables 3.6 and 3.7. Therefore, a three-person group will secede and form a separate three-person coalition if its members receive less than 15 units of value in the four-person coalition. This means that any possible three-person group must receive a total of 15 units of value. To accomplish that, each individual must receive 5 units of value. But that adds up to 20 units of value, and the four-person coalition can produce only 18. Thus the four-person coalition cannot pay out enough to prevent some three-person group from seceding: the grand four-person coalition cannot be stable.

Table 3.8 Yet another Four-person Cooperative Game

Coalition of Size	Value
1	1
2	8
3	15
4	18

But the three-person coalitions are not stable either. A three-person coalition will leave the fourth person in a singleton coalition with a value of only one. That individual can approach one of the members of the three-person coalition and offer to form a two-person coalition with him. The singleton might offer to settle for 2, leaving a payoff of 6 for the individual who defects from the three-person coalition. (Remember, equal payoffs are not required, and in this case inequality is an advantage to the singleton, even though he receives less than the other player, since he still does better than he can do as a singleton.) Both are better off, so the three-person coalition is not stable either. We have seen, in passing, that a one-person coalition also is not stable. But the two-person coalition is not stable either, since a merger of two two-person coalitions (or of one two-person and two one-person coalitions) can again make all better off. Starting from two two-person coalitions, with payoff schedules 6,2 in each coalition, forming a four-person coalition allows the addition of ½ to each individual's payoffs, with a payoff schedule of 6½, 6½, 2½, 2½. (Remember, equality is not required.) But this arrangement is not stable either . . . and that is the point. In this game, there are no stable coalitions and payoff schedules. The core of this game is an empty set, which is to say that there are no coalitions with payoff schedules that fulfill the stability standards set by the core of a cooperative game.

We have seen that the core has two mathematical shortcomings as a solution to a cooperative game. First, there may be many solutions to some games, and second, some games may have no solutions at all. Put otherwise, depending on our assumptions, there may be just one stable arrangement consistent with the competitive alternatives, or many, or none. A traditional (neoclassical) view in economics is that neither of the latter two possibilities is real, and that there is always just one stable arrangement in the economy. Of course, neoclassical economics makes different assumptions than we see in these examples, and in some ways those assumptions are more realistic. To begin with, the world is much bigger than our examples, and in the real world there will be many coalitions competing (non-cooperatively) with one another. Second, even in the *short run* – that is, with the population sorted into coalitions that do not change – there will be ways for each coalition to increase its profits at the expense of another. For example, it may be possible to increase the value created, at the margin, by using more raw materials along with the labor supplied by the existing members. In neoclassical economics, we find that these marginal adjustments will reduce the stable payoff schedules to just one. In the *long run* – that is, allowing people to shift to new coalitions and allowing coalitions to shift their use of capital goods – there will also be just one stable solution. Thus, when we allow for higher order goods (raw materials and capital goods) and a much larger number of players and coalitions, it is quite possible that these difficulties may disappear. On the other hand, the assumptions of neoclassical economics are no more sacrosanct than those we have made in this section, and in future chapters we will see that the problems we have seen here can occur in more realistic examples.

Suppose, then, that there are many payment schedules that can be stable against competitive alternatives. How are we to determine which payment schedule will be chosen? Evidently this is a matter of bargaining power. Consider, again, the game in Table 3.6. For stability, at least 16 units of value must be distributed equally among the four members of the four-person coalition. The surplus over this amount is 2 units. If the four agents have equal bargaining power, presumably they will divide the 2 units equally. If they have unequal bargaining power this will be reflected in an unequal distribution of the surplus of 2.

Unfortunately but truthfully, we know very little about bargaining power. The study of bargaining in modern economics began with the Danish economist, Frederik Zeuthen.(13) Zeuthen proposed that a bargainer would decide whether to accept or reject an offer by balancing the risk of a breakdown of bargaining, resulting in no deal, against the benefits of the offer itself. Bargaining would come to an end at a deal in which the risk would just balance the benefits. John Nash also did a study of bargaining.(14) In a sense, he rediscovered Zeuthen's model and his reasoning about balancing risks against benefits, but he added two very important insights. First, *any* bargain that distributes the entire surplus between the two bargainers could be a Nash equilibrium, and so it would be rational in that limited sense. However, Nash also showed that the bargaining solution Zeuthen had proposed could be derived from a small number of assumptions that could be seen as characterizing a reasonable compromise between the interests of the two bargainers. This theory is widely called the Zeuthen-Nash bargaining theory and has often been applied in economics. However, it has three shortcomings. First, it only applies to bargaining between two parties. Despite a number of proposals, there is no consensus on how bargaining may affect the distribution of the benefits of a coalition between more than two parties to the distribution. Second, it often does not agree well with the evidence. Third, it does not allow for any given differences in the bargaining power of the bargainers. When the Zeuthen-Nash theory is modified to allow for difference in the bargaining power of the adversaries, it fits the evidence better, but that raises a question: just where do these differences in bargaining power come from? Some applications of bargaining power in economics do use the modified Zeuthen-Nash bargaining model with given differences in bargaining power, sometimes with some success.

This is not to say that observable conditions have no impact on bargaining power. For example, if the institutional system encourages or requires collective bargaining between employers and employees, this can shift the bargaining power from employers to employees. In any case, we have a great deal to learn about bargaining power. Without attempting here to state a general theory of bargaining power,[15] the chapters to follow will sometimes apply the following assumptions: bargaining power arises from threats. In a successful cooperative arrangement the threats are not carried out, so are not necessarily observable. Instead, rational bargainers each assess the threats that others may make and moderate their demands so that a mutual agreement can occur. When we observe threats carried out, the attempt to form a cooperative coalition has failed. The threats may be of two kinds: those that, in Nash's words, "will not be something [the threatener] would want to do, just for itself" (i.e., that can be carried out only at some cost) and those that are. Threats of the latter sort may be expressed as rationality constraints, if they require the threatening group to withdraw from the coalition. However, there will be a residual of threats of both kinds, either threats that the threateners cannot carry out without regret or that can be carried out without dissolving the coalition, which will influence the bargaining power of one bargainer or another among rational bargainers.

COOPETITION

In game theory, the distinction between cooperative and non-cooperative games is often treated as absolute, as the two approaches reflect different, simplified conceptions of rational decision-making. In the world of our experience, human action seems to be a combination of cooperative and non-cooperative activity. That is, in the real world we observe "co-opetition," to borrow the title of a book by Brandenburger and Nalebuff.[16] As always, more realism means less simplicity, and there are a few studies in game theory and economics that have tried to deal with this combination of cooperative and non-cooperative decision-making in ways that are complex, deeply mathematical, or both. (I have written on this subject myself.)[17]

In the real economy we observe a remarkable mixture of cooperative and non-cooperative decisions and arrangements. As we will see in the next chapter, exchange, which is inherently a cooperative arrangement, is central to modern economies and is another source of the benefits of working together, and yet markets, the principle institution for the organization of exchange, are always thought of as non-cooperative. Is this not a puzzle? It would seem that a major objective of economists should be to understand this mixture of cooperative and non-cooperative action. This goal will in any case be central in the chapters to follow in this book.

CHAPTER SUMMARY

In a modern economy, the actions of people are highly interdependent, and as a consequence, their decisions are interdependent. Game theory, despite its name, can be thought of as the study of interdependent decisions. There are two great branches of game theory: non-cooperative and cooperative game theory. Non-cooperative game theory alerts us to some deep problems of interdependent decisions and therefore of interdependent actions. For this purpose, Nash equilibrium is a key tool. In a Nash equilibrium, people may rationalize their decisions in sophisticated or simple ways, but since they all choose the best responses to the decisions of others, their rationalizations will be verified by their experience. We see that in some cases, rational and self-interested decisions may make the decision-makers jointly worse off than they would be if they were less rational. Other kinds of problems arise when there are two or more Nash equilibria. In each case we may contrast the non-cooperative with a cooperative solution, using ideas about efficiency, common in economics, to characterize the cooperative solution. We may say that the cooperative arrangements create value, in that they enable the participants to obtain a more productive, profitable or satisfactory result than they can get in a non-cooperative interaction. But this value can often be distributed among the participants in the cooperative coalition in many different ways. Competitive alternatives and bargaining power together determine the division of the gains, but unfortunately we have little understanding of bargaining power. Further, in the actual world, we observe a mixture of cooperative and non-cooperative decision-making and action. An objective of this book is to better understand this mixture.

SOURCES AND READING

Game theory is introduced in several textbooks. Naturally I prefer and recommend my own: Roger A. McCain (2014), *Game Theory: A Nontechnical Introduction to the Analysis of Strategy*, 3rd edition, Singapore and Hackensack, NJ: World Scientific. (1) On Schelling and Aumann, see Chapter 1, note 8. (2) On George Poulett Scrope, see note 5, Chapter 2 in this volume. (3) John von Neumann and Oskar Morgenstern (2004), *Theory of Games and Economic Behavior*, Sixtieth Anniversary Edition, Princeton, NJ: Princeton University Press, originally published 1944. (4) The discussion of the history of the Prisoner's Dilemma and the early history of game theory largely follows William Poundstone (1992), *Prisoner's Dilemma*, New York: Doubleday. (5) Nash published his ideas in two research papers, John Nash (1950), "Equilibrium points in n-person games," *Proceedings of the National Academy of Science*, **36**(1), 48–9; and John Nash (1951), "Non-cooperative games," *Annals of Mathematics*, **54**(2) (Sept.), 286–95. The subsequent literature is enormous. (6) The difficulties that arise from the existence of two or more Nash equilibria were discussed in R. Duncan Luce and Howard Raiffa (1957), *Games and Decisions*, New York: Wiley and Sons, and some subsequent literature. (7) The concept can be traced to two research papers, B. Douglas Bernheim (1984), "Rationalizable strategic behavior," *Econometrica*, **52**(4) (July), 1007–28, and David G. Pearce (1984), "Rationalizable strategic behavior and the problem of perfection," *Econometrica*, **52**(4) (July), 1029–50. There is a modest subsequent literature.

(8) The best source for Pareto is probably Vilfredo Pareto (1971), *Manual of Political Economy*, New York: A.M. Kelley. Pareto optimality became a standard concept in economics with the development of "the new welfare economics." In this the work of Sir John Hicks was crucial. See especially John R. Hicks (1939), "Foundations of welfare economics," *Economic Journal*, **49**(4) (Dec.), 696–712.

(9) See, for example, John Bates Clark (1899), *The Distribution of Wealth*, New York: Macmillan, p. 3. (10) The word "strategy" is used in several different senses in game theory. In the example in this section, the side payment is more precisely a "behavior strategy," as Kuhn used the term in Harold W. Kuhn (1953), "Extensive games and the problem of information," in Harold W. Kuhn and Albert W. Tucker (eds), *Contributions to the Theory of Games, Vol. II Annals of Mathematics Studies, No. 28*, Princeton, NJ: Princeton University Press, pp. 193–216. This is probably the most common interpretation in the literature of game theory, and I will try to use it consistently in this sense in the book. (11) The word "cooperative" is used in a different sense here, not in the sense of cooperative game theory but in the sense of the international cooperative movement. See International Cooperative Alliance (1995), "Statement on the co-operative identity," available at http://www.ica.coop/en/what-co-op/cooperative-identity-values-principles, as of 25 July 2013. (12) See "Mondragon," available at http://mondragon-corporation.com/ENG.aspx as of 25 July 2013; and Baleren Bakaikoa, Anjel Errasti and Agurtzane Begiristain (2004), "Governance of the Mondragon Corporación Cooperativa," *Annals of Public and Cooperative Economics*, **75**(1), 61–87, among many other studies.

(13) Zeuthen's bargaining theory was published in his 1930 book, *Problems of Monopoly and Economic Warfare*, London: Routledge and Kegan Paul. (14) Nash wrote two papers on bargaining: John Nash (1950), "The bargaining problem," *Econometrica*, **18**(2), 155–62; and

John Nash (1953), "Two-person cooperative games," *Econometrica*, **21**(1) (Jan.), 128–40. The discussion of Nash's bargaining theory draws on both papers. (15) A theory that attributes variable bargaining power to subgroups of all sizes in games of more than two persons is developed in Chapter 6 of Roger A. McCain (2013), *Value Solutions in Cooperative Games*, Singapore and Hackensack, NJ: World Scientific.

(16) The term "coopetition" is from Adam Brandenburger and Barry J. Nalebuff (1997), *Co-opetition: 1. A Revolutionary Mindset that Combines Competition and Cooperation; 2. The Game Theory Strategy that's Changing the Game of Business*, New York: Doubleday. With Harborne Stuart, Brandenburger has proposed a general model of business strategy that combines cooperative and non-cooperative elements, which they term "biform games." On this see Adam M. Brandenburger and Harborne W. Stuart Jr. (1996), "Value-based business strategy," *Journal of Economics and Management Strategy*, **5**(1) (Spring), 5–24; Adam Brandenburger and Harborne W. Stuart Jr. (2007), "Biform games," *Management Science*, **53**(4) (Apr.), 537–49. (17) For my writing on games that are partly cooperative and partly non-cooperative, see Roger A. McCain (2009), *Game Theory and Public Policy*, Cheltenham, UK and Northampton, MA, USA: Edward Elgar.

4. Exchange: how difference enriches us

In this part of the book, we are concerned with the relation between affluence and interdependence. An old saying has it that "fair exchange is no robbery," and in fact, exchange, per se, is a form of interdependence that can make both sides better off. Of course, common sense has known this for centuries, but it was not until the late 1800s that economists developed their modern theory of exchange. The problem is that in exchange, per se, the benefits are subjective. Each person has something he or she likes better than what he or she gave up, but "likes better" is a subjective state of mind. How can we base our theory on subjective states of mind? The classical economists, from Smith through Mill, would not take this step, but in the later 1800s, the founders of neoclassical and Austrian economics did take the step of incorporating subjective states of mind in their new theory of exchange.

On the one hand, trade has existed in human society for a very long time. We have evidence of stone tools made from stone formations in Switzerland that have been found as widely in the old world as Eastern Siberia and South Africa. Presumably they got there by exchange – exported from Switzerland! On the other hand, exchange played very little part in the daily lives of most people. Walter Neale's description of life in a pre-industrial Indian village illustrates this.(1) The village was self-sufficient: people lived on what they themselves or their neighbors grew and made. There was interdependence within the village, but exchange played little role in it. Personal relations and customary practices, not prices and exchange, were the framework for the villagers' interdependence. A portion of their harvest was taken in tribute by a distant monarch. The monarch, no doubt, relied on exchange and trade over large distances for many of his luxury goods, the tools for the craftsmen in his household, and the weapons his soldiers used to collect the tribute. In a modern society, by contrast, we all rely on huge numbers of people we do not know, some very distant, to provide us with the simplest goods and services. For these impersonal interdependencies, markets and exchange are necessary and play a key part in our lives almost every day. This is characteristic of our affluent modern societies. Exchange is well understood in the economics of the twentieth century and accordingly this chapter will be very conventional in important ways.

USE VALUE AND EXCHANGE VALUE

Once again we may begin with Smith. In the work that founded economics, *The Wealth of Nations*, Adam Smith[(2)] certainly made the central importance of exchange very clear. In a famous phrase, he observed that interdependency and affluence were "the necessary, though very slow and gradual consequence[s] of a certain propensity in human nature which has in view no such extensive utility; the propensity to truck, barter, and exchange one thing for another." But this understanding led at once to a troubling puzzle. Common sense suggested that the usual or normal price of a good would depend upon its usefulness, its utility. However, the facts did not conform to this common sense idea. This was particularly obvious in a contrast between the normal prices of diamonds and water, and so the puzzle came to be known as the "Paradox of Diamonds and Water." Smith wrote:

> Nothing is more useful than water: but it will purchase scarce anything; scarce anything can be had in exchange for it. A diamond, on the contrary, has scarce any value in use; but a very great quantity of other goods may frequently be had in exchange for it.

This led Smith to believe that there is no relation between usefulness and price, and he distinguished value in use (use value) from value in exchange (exchange value). If not usefulness then, what might determine value in exchange? Smith began to develop a theory to explain that – a theory that became known as the labor theory of value. For most of a century, economists worked to develop that theory, but we shall not follow up their work here, because a new solution was found in the later nineteenth century, and that new solution transformed economics.

UTILITY AND WILLINGNESS TO PAY

A key insight had come from Thomas Malthus. Suppose we have only a limited quantity of some item – let us say, for example, very rare signed editions of a fine book. Let us suppose that just 40 copies are available, but there are 50 collectors who would like to have a copy. Malthus wrote:[(3)]

> Let us suppose a commodity in great request by fifty people, but of which, from some failure in its production, there is only sufficient to supply forty. If the fortieth man from the top have two shillings which he can spend in this commodity, and the thirty nine above him, more, in various proportions, and the ten below, all less, the actual price of the article, according to the genuine principles of trade, will be two shillings. If more be asked, the whole will not be sold, because there are only forty who have as much as two shillings to spend in the article; and there is no reason for asking less, because the whole may be disposed of at that sum.

There are two vitally important insights here. One is that the price of an item in limited supply will depend on the willingness of potential buyers to pay for it, and the other is that the price will correspond to the willingness to pay of the one who is least willing, but at the same time will pay enough that he will be able to buy. This buyer – the fortieth, in Malthus's story – would be called the marginal buyer in modern economics, and his or her willingness to pay is the marginal willingness to pay. So the price of our collectable fine book will depend on the marginal willingness to pay.

But this is only a beginning, since we (and Malthus) have simplified the problem by assuming that each buyer buys only one copy of the book. For many kinds of goods and services, from shirts to songs downloaded from an online music store, a buyer may often buy more than one. What determines how many, and how does this relate to the price and the marginal willingness to pay? To get the answer to that we need to go back to the other side of Smith's paradox of value, utility.

Utilitarianism

Utility was at the center of the ideas of a very controversial philosopher, Jeremy Bentham.(4) Bentham argued that laws and moral rules should be based on utility. For an individual, Bentham said, utility would be the sum total of pleasure the individual enjoys, minus the pain he experiences – pleasure minus pain. Laws then should be adjusted so as to increase the total utility of the entire population. Good moral rules would be rules that would lead to the greatest utility for the whole population. Utilitarianism is also a theory of motivation. In that application it is assumed that people ordinarily, when they are not acting according to moral rules, choose the course of action that makes their own utility as great as possible.

Utilitarianism has something of an undeserved bad reputation. The idea that moral rules, laws, and public policy should be determined by pleasure-seeking seems to some to be low and wrong. But, in fact, utilitarianism is a very demanding moral standard, and some moral philosophers have criticized it as being too demanding. To take a simple example, suppose I have to decide whether to buy cheap coffee, or a certified fair trade coffee that is more expensive because the poor people who raise the coffee beans and produce the coffee are better paid. In making that decision, a utilitarian moralist would balance the utility he or she gives up from paying the higher price against the utility the poor producers of the fair trade coffee would get from the higher price. The utilitarian might well feel morally compelled to buy the fair trade coffee. In general, utilitarian moral rules would require us to consider the consequences for the utility of *every* human being affected by our decision. And not only human beings. The pains suffered by animals must also be taken into account. All beings should have their pains and pleasures considered *equally.* This is indeed a difficult moral standard.

As a standard for law and public policy, the utilitarian would say that a good law or a good policy is one that does someone some good, and which does the most good overall. But, for the utilitarian, to "do someone good" is to create a situation in which the someone enjoys more good subjective states of mind and fewer bad ones – more "pleasures" or less "pains." That does not seem an impossible standard to meet (though it might require a great deal of information) and it is a simple standard – but perhaps too simple.

Bentham was about a generation younger than Smith, but senior to Malthus. Bentham's first book appeared in the same year as *The Wealth of Nations*, but he did not influence either Smith or Malthus. However, some of the followers of Smith and Malthus were also followers of Bentham. Among them was James Mill, a noted economist in his time, and roughly a contemporary of Malthus. James Mill's son, John Stuart Mill, was one of the great economists and philosophers of the nineteenth century, and, after an emotional crisis as a young man, John Stuart developed his father's ideas, both on economics and utilitarianism, though perhaps not in directions has father would have liked.(5) John Stuart Mill restated and refined Bentham's utilitarianism, allowing more for human ignorance and fallibility and for the subtle differences of quality among the subjective states of mind we refer to as "pleasures" and "pains." His ideas were an extraordinary (perhaps impossible) blend of the ideas of Smith and Malthus, utilitarianism, Christian socialism, and a passionate commitment to liberty and to the equality of the sexes. It took a truly great mind to reconcile these very diverse elements – and many twentieth-century critics have doubted that even Mill was actually able to do it. But there was little doubt in his own lifetime, and Mill was the most influential economist and social philosopher of the nineteenth century, with the *possible* exception of Karl Marx.

So, James and John Stuart Mill brought utilitarianism and economics together, but there was no real marriage of the two. For the Mills, economics and utilitarianism remained separate ideas they drew on for social and political philosophy. Nevertheless, among the followers of the Mills, the idea emerged that economics ought to be *based* on utilitarianism. The idea is that economics should begin with the utilitarian theory of *motivation*, that people would act in the ways calculated to increase their utility. William Stanley Jevons[(6)] was the first to propose that idea, and he was followed independently by several other economists. Jevons wrote: "A true theory of economy can only be attained by going back to the great springs of human action – the feelings of pleasure and pain." Thus, Jevons adopted the utilitarian theory of motivation. This new idea in economics had a wide influence because it enabled economists to do what Smith could not: explain the Paradox of Diamonds and Water. (And that is the reason for this long-winded digression on utilitarianism.)

A Utilitarian Theory of Exchange

Let us see how a utilitarian explains the Paradox of Diamonds and Water. Consider John Doe, who has several shirts in his drawer, which he can wear on different days. If he has only one shirt, this meets his basic need for warmth and decency, and he will get a very large amount of utility from that. With a second shirt, he can wear one while he washes the other, and that will increase his utility by a pretty large number, too. However, if he has several shirts, an additional shirt will add only a little to his utility. To be specific, suppose he has eight shirts in his drawer. He can wear one of them while he takes the rest to the laundromat once a week, and can wear a different shirt every day. If he now buys a ninth shirt, that will increase his utility somewhat – he will enjoy even more variety in the shirts he wears from day to day, particularly if it is a nice shirt, and as each shirt is worn a little less often they all will last a little longer – but the more shirts he has, the less impact an additional shirt will have on his total utility. Thus, we see that his total utility increases as he has more shirts, but it increases at a decreasing rate. We define the marginal utility of shirts as the additional utility that John Doe will obtain as the result of adding *one* additional shirt to his drawer.

In general, we define the marginal utility of any good or service as the utility of one additional unit of that good. Marginal utility, like total utility, will vary as the quantity that the individual consumes varies. When one quantity (utility) varies along with another (consumption) in this way, a mathematician would say that the first quantity is a function of the second, and economists say that total and marginal utility are functions of the quantity consumed, and speak of the utility function and the marginal utility function. But, as we have seen, the total utility of the good will increase at a decreasing rate. Marginal utility corresponds to the rate of change* of total utility, so we may say that, beyond some quantity of consumption, the marginal utility of any good or service will decrease. Since the late nineteenth century this has been called the law of decreasing marginal utility. This law is a fundamental principle of economics.

How does this resolve the Paradox of Diamonds and Water? The key to it is that exchange value depends on the marginal utility of the good, while value in use corresponds to the total or average utility, which is quite different. Water was very plentiful in Britain, a wet country, in 1776. Because it was very plentiful, its marginal utility was very small, even though its total and therefore its average utility was great. Diamonds were very scarce, and so their marginal value was quite high, so their exchange value was quite high. In general, because of the law of decreasing marginal utility, scarce goods would command higher exchange values, relative to use values, than goods that are plentiful. Indeed, this solution resolves the paradox in such a satisfactory way that in the twentieth century few economists (other than Marxists) continued to talk about exchange value and use value at all. Like Malthus, we focus on the marginal willingness to pay, which is determined by the marginal utility of the good or service.

Moreover, to close the circle, the law of diminishing marginal utility helps us to understand exchange. Here, we will illustrate that with an example adapted from the writing of Carl Menger.(7) For the example, we consider two frontiersmen, one of whom possesses ten horses, and the other possesses ten cows. Their total and marginal utility of horses and cows are shown in Table 4.1.

At the beginning of the example, the First Frontiersman has ten horses and the Second Frontiersman possesses ten cows. Now suppose that they exchange one horse for one cow. This increases the utility of the First Frontiersman by 9 units, from 55 to 64. This improvement of 9 is equal to the difference between the marginal utility of the first cow, 10, and that of the tenth horse, 1. Similarly it increases the utility of the Second Frontiersman by 9. Now suppose that they exchange a second horse for a second cow. This increases the utility of each of them by 7 units, that is, 9 (the marginal utility of the second horse or cow) minus 2 (the marginal utility of the ninth horse or cow). A third exchange will increase their utility by 5, and the fourth by 3. A final exchange will increase the utility of

* This can be expressed in terms of calculus, and for some purposes that is very convenient – hence the pages of mathematics in many economics books, including some of mine. Here "rate of change" would be more precisely expressed by saying that the marginal utility of a particular commodity is the partial derivative of the utility function with respect to the quantity consumed of that commodity.

Table 4.1 Utility of domestic animals

	First Frontiersman				Second Frontiersman			
Number	Total utility of horses	Marginal utility of horses	Total utility of cows	Marginal utility of cows	Total utility of horses	Marginal utility of horses	Total utility of cows	Marginal utility of cows
1	10	10	10	10	10	10	10	10
2	19	9	19	9	19	9	19	9
3	27	8	27	8	27	8	27	8
4	34	7	34	7	34	7	34	7
5	40	6	40	6	40	6	40	6
6	45	5	45	5	45	5	45	5
7	49	4	49	4	49	4	49	4
8	52	3	52	3	52	3	52	3
9	54	2	54	2	54	2	54	2
10	55	1	55	1	55	1	55	1

each frontiersman by 1, leaving each of them with five horses and five cows and utility of 80.

In this example we have assumed that the two frontiersmen have identical utility functions – which is to say that their tastes and preferences for cows and horses are the same – although they begin with different endowments, that is, with different stocks of cows and horses. Now let us reverse those assumptions, and assume instead that they have different tastes – that the First Frontiersman really likes cows and the Second Frontiersman really likes horses. Their total and marginal utilities are shown in Table 4.2.

At the beginning of the example, they each have five horses and five cows, so we would say, in the jargon, that their endowments are alike, although their tastes, as expressed by their utility functions, are different. Thus, they each begin with utility of 130. Now suppose that they exchange one horse for one cow, with the First Frontiersman giving up one horse and the Second Frontiersman giving up one cow. This will increase the utility of the First Frontiersman by $15-6=9$, the difference between the marginal utility of the sixth cow (gained) and the fifth horse (given up) and it increases the utility of the Second Frontiersman by the same 9 units. In this way they will continue to exchange until the First Frontiersman has all the cows, for a total utility of 155, and the Second Frontiersman has all the horses, with a total utility also of 155.

What we see in both of these examples is the following:

- Exchange can make both parties to the exchange better off.
- This will be so if they have the same tastes (utility functions) but different endowments.
- And will also be true if they have different tastes and the same endowments.
- But, on the other hand, if their tastes and endowments are the same (as in Table 4.1 when each has five horses and five cows) then further exchange will do no good and will make them worse off.

More generally, what we see here is that by means of exchange, *difference enriches us*. In this example, interdependence takes the specific form of exchange, and, with differences either of endowments or of taste, interdependence is the necessary condition for the enrichment.

Beyond Utilitarianism?

In this section, we have adopted the utilitarian view that the subjective satisfactions that people get from their consumption and other activities,

Table 4.2 Utility of domestic animals with different tastes

	First Frontiersman				Second Frontiersman			
Number	Total utility of horses	Marginal utility of horses	Total utility of cows	Marginal utility of cows	Total utility of horses	Marginal utility of horses	Total utility of cows	Marginal utility of cows
1	10	10	20	20	20	20	10	10
2	19	9	39	19	39	19	19	9
3	27	8	57	18	57	18	27	8
4	34	7	74	17	74	17	34	7
5	40	6	90	16	90	16	40	6
6	45	5	105	15	105	15	45	5
7	49	4	119	14	119	14	49	4
8	52	3	132	13	132	13	52	3
9	54	2	144	12	144	12	54	2
10	55	1	155	11	155	11	55	1

allowing also for their dissatisfactions, discomforts, pains, and sorrows, can all be expressed in summary terms as a number that can be compared, added, and subtracted for the different people in society. But this has been a controversial idea among economists and philosophers. Some economists and philosophers continue to defend it, while others reject it without reservation. We shall have to reserve judgment on that. Twentieth-century economics developed a slightly different approach to the theory of exchange, one that is a bit more cautious in making simple assumptions about motivation. This has a further advantage in that we can take realistic account of some complexities about the relations among consumption goods.

We observe in passing that if two people have the same utility functions, it is the same thing as saying that they have the same tastes or preferences. Where Jevons and Menger had proposed to explain the advantages of exchange in terms of utility, many later economists have followed the example of Francis Ysidoro Edgeworth(8) and explained this in terms of preference. The idea is that a person confronted with a choice among a number of alternatives will choose the one she or he most prefers.

This is still a kind of utilitarianism – philosophers call it preference utilitarianism – but it has the advantage that it doesn't require a numerical utility quantity that we cannot measure directly, and perhaps cannot measure at all. On the other hand, a little introspection tells us that preferences *are* among our subjective experiences, and very few people – perhaps a few religious mystics – would deny that they have similar experiences. More than that, we can *observe* people choosing one thing rather than another, and if they are rational in the appropriate sense, this can show us just what they prefer. Choices give us evidence on preference. We often say that the rational person's choices, in some circumstances, "reveal" his or her preferences and this is a theory of revealed preference.(9) Surprisingly, perhaps, the discussion of preferences can get quite technical and mathematical. Many of us would say that we prefer coffee to tea, or vice versa, but that is not quite what economists mean. For purposes of economics, preferences are usually expressed as an ordering over combinations of quantities of consumer goods and services.

These are familiar ideas from the economics of the twentieth century, and still central to advanced microeconomic theory. However, for the purposes of this book, they need little further discussion. Most of the discussion to follow can be expressed in terms of preferences rather than utilities or money payments, with little modification of the results but a great deal more technical difficulty.

Interim Summary

Whether we accept the utilitarian approach or not, it will serve as an illustration of the fundamental point central to this chapter: exchange is another way that people can make themselves better off by working together. In the case of exchange, objects such as horses and cattle are reallocated from those who desire them less to those who desire them more. In the language of twentieth-century economics, exchange improves the allocation of resources. The further insight we gain from the theory of exchange is that exchange can make us better only if the exchange is between people who differ – who have different endowments or different tastes or both. Adam Smith already knew that. People are different for many reasons, but, as Smith alerts us, a major source of difference is the different roles that they play in the division of labor. As he wrote:(10)

> Every workman has a great quantity of his own work to dispose of beyond what he himself has occasion for; and every other workman being exactly in the same situation, he is enabled to exchange a great quantity of his own goods for a great quantity, or, what comes to the same thing, for the price of a great quantity of theirs.

This is an insight we ought not lose sight of in the twenty-first century.

EXCHANGE AS COOPERATION

In the previous chapter we observed that interdependencies can be modeled as games, leading to a subdiscipline called game theory, but that there is a crucial distinction between games that are played non-cooperatively and those that are played cooperatively. In a cooperative game, decisions are made with a view to mutual benefit, while in a non-cooperative game, decisions are made with a view to the individual's benefit, while the individual regards decisions of the other players as given circumstances to which the individual must adapt. We also observed that the actual world seems to show us a mixture of cooperative and non-cooperative decision-making. Now here is a key application: exchange is a cooperative game.

To illustrate this point, let us return to Menger's frontiersmen and their horses and cows. As a first step, let us try to think of their exchange as a non-cooperative game. What strategies does each person have? The First Frontiersman, acting unilaterally and taking the other's decisions as givens that he cannot influence, can give up one, two, three, . . . or ten horses, or none. He cannot, *unilaterally*, obtain any cows. Similarly, the Second Frontiersman can give up one, . . . ten cows, or none, but cannot, unilaterally, obtain any horses. To keep things simple let us consider only two strategies: give up five horses or none, and give up five cows or none. This gives us a game as shown in Table 4.3.

Table 4.3 A simplified Exchange Game

First Payoff to First Frontiersman, Second to Second Frontiersman		Second Frontiersman	
		Give five	Give none
First Frontiersman	Give five	80,80	40,95
	Give none	95,40	55,55

This game has a dominant strategy, and that is to give none. Thus, in a non-cooperative framework there will be no exchange. (This follows the same way in a larger game in which we consider all 11 strategies for each frontiersman: give none will always be a dominant strategy.) On the other hand, the strategies "Give five, Give five," yield the largest total value at 160, and in that sense, appears to correspond to a cooperative solution. It need not be the only cooperative solution. If the two frontiersmen have equal bargaining power, then this equal distribution of the gains from trade will be the cooperative solution. If the First Frontiersman is the stronger bargainer, then the Second Frontiersman might have to put up something more – a few measures of grain for fodder for the animals, perhaps – in order to close the deal. But the key point for our purposes is that there is no exchange in a non-cooperative game; exchange is a cooperative phenomenon.

Indeed, in the 1960s and 1970s, the cooperative approach to exchange was highly developed.[11] Cooperative game theorists adapted Edgeworth's model of exchange based on preferences, and perfected and extended it. They showed how increasing the number of traders could limit the influence of bargaining power on the terms of the exchange, increasingly centering on a particular rate of exchange that can be identified with what twentieth-century economics called the market equilibrium price. This line of research seems since to have been abandoned, probably because it proved difficult to extend it to a world in which goods must be produced before they are exchanged and in which some people may not have crucial information that they need to make rational decisions. It is important work all the same, and not only because it points up a key fact for this chapter: that exchange is a cooperative phenomenon, and could not occur in a purely non-cooperative world.

AUCTIONS AND MARKETS

Menger's examples are simplified in that there are only two frontiersmen, isolated in some remote district, to engage in exchange. Economists are sometimes ridiculed for such simplified examples, and not very justly – simplification is often a good first step. But Malthus's example is a little more realistic in this sense: there will often be more than just two people who might enter into exchange. Moreover, as we have seen, exchange is an instance of cooperative action in the puzzling mixture of cooperation and non-cooperation that defines our economic existence. One way to organize exchange is through an auction. The interplay of cooperation and non-cooperation can also be illustrated by auctions, and auctions are important in themselves, so this section will introduce some economics and game theory of auctions.

For example, we suppose that Agent 1 owns a unique Old Master painting, but would prefer to have money instead (if she can get enough). Agents 2, 3, . . . N all have some money and would like to buy the painting (if the price is right). Each of the agents 2, 3, . . . N has a reservation price p_i, which is the value that Agent i sets on the painting. That is, if Agent i can buy the painting for less than p_i, then she will be better off (in terms of her own subjective utility or preferences) with the painting than with the money. Put otherwise, Agent i will get more good states of mind from possessing the painting than from anything else that she can buy for less than p_i. If the price of the painting is more than p_i then the opposite is true. Agent 1 also has a reservation price, p_1, which is the cheapest price she can accept and be no worse off. We assume that $p_1 < p_i$, for at least some *i*, since otherwise Agent 1 would not be offering the painting for sale. Other than that, each person knows her own reservation price but does not know that of any other potential buyer.

Since Agent 1 wants to realize the money value of her antique painting, she will hold an auction. The auction will proceed as follows: interested buyers will gather at one location where an auctioneer will call out possible prices. The auctioneer will begin with p_1, the reservation price for the sale,* and raise the price in small steps. At each step those who are willing to pay that price raise their hands and keep the hands up until they judge the price to be too high. The auction stops when only one person has his or her hand raised and he or she buys the painting for the last price.(12)

Think of the bidding in the auction as a non-cooperative game. The

* It is common in an auction to set a minimum price, called the reservation price for the auction, and this is where economists derived the term reservation price for a minimum or maximum acceptable price in general.

strategy for each bidder is the price at which he or she will drop out of the bidding, taking her hand down. There is a dominant strategy in this game, and the dominant strategy is to continue in the bidding until the auctioneer calls out the bidder's reservation price, and to drop out at that point. That way, if she buys the painting she is better off for it, and if she does not buy it she is no worse. Any shift away from this strategy may make her worse off (depending on what others do) but cannot make her better off.

Thus, we have a unique Nash equilibrium for the bidding game, in which each bidder keeps her hand up so long as the auction price is below her reservation price. As a result, the person who buys the Old Master will be the buyer with the highest reservation price, and she will pay the second-highest reservation price. This is an efficient Nash equilibrium. In an auction of this kind it is efficient for the buyer with the greatest reservation price to buy the item. (We will see just below why this is so.)

We have treated the bidding in the auction as a non-cooperative game – but the participants in the auction itself are the members of a cooperative coalition, and the conduct of the auction is a cooperative joint action. Each person who comes to the auction to bid makes a decision to do so: to be present at the time and site of the auction.* Each one presumably makes this decision because she sees a possibility of benefiting from it.

For convenience, let Agent N be the one with the greatest reservation price, N–1 the second highest, and so on, so that $p_N \geq p_{N-1} \geq \ldots \geq p_2$. The value of the Old Master painting from the point of view of Agent N is p_N and the sacrifice by Agent 1 in giving up the painting is p_1, so the surplus created by the auction coalition is $p_N - p_1$. If the item were sold to any other agent, for example Agent *i*, then the value created would be $p_i - p_1$, and since $p_N > p_i$, this value would be less than $p_N - p_1$. Thus, it is efficient to sell the item to the bidder with the highest reservation price, in that this sale maximizes the value created by the auction coalition. The auction also determines the division of this surplus, as Agent N gets a net benefit of approximately $p_N - p_{N-1}$ and the seller gets about $p_{N-1} - p_1$.**

Apart from Agent N and the seller, no one gets any benefit from the auction of the Old Master. Why then should they participate in the auction at all? Remember that before the auction takes place none of the buyers knows the reservation prices of the others. Thus, each person

* Online auctions, which have become important since the late 1990s, are not an exception, nor are sealed bid auctions in which the buyers need not come together in the same room. In each case a number of potential buyers must make decisions that are coordinated with one another and with the seller's decision: for example, the decision to direct a browser application to the web address for the seller.

** Here is a complication: if $p_{N-1} < p_1$, then the sale must be at p_1, since only Agent N will raise her hand at the minimum sale price, and in that case Agent N gets $p_N - p_1$ and the seller gets nothing from the surplus created by the sale.

sees some possibility that she will be Agent N and will be the one who benefits by buying the Old Master. That is why Agents 2 through N − 1 participate. Their decisions are made despite their uncertainty as to whether they will benefit or not. We will take up decisions made under uncertainty in the next chapter.

Nevertheless, there may be a problem here for the organizer of the auction. If the auction is to be successful, it must attract the participation of a number of potential buyers. For a real Old Master – a Rembrandt, Tintoretto, or Rubens, for example – this would probably be little problem. Nevertheless, some important, high-profile auctions have failed for lack of participation.(13) This is probably a reason that real auctions often sell a number of different items or "lots" that could appeal to many of the same buyers. The more lots are sold, the better the chance that a particular buyer will buy at least one of them and so share in the coalition surplus, and so the likelier that person would be to participate.

Thus, the bidding process is non-cooperative, but the organization of the auction is cooperative, and the seller and the bidders are participants in a cooperative coalition. The rules of the non-cooperative game have been carefully designed to realize the maximum total surplus for the buyer and seller. The non-cooperative bidding process has some advantages. It does not require much information or calculation – each bidder needs to know only her reservation price. It also resolves any issues of bargaining power.

However, cooperative behavior can influence the auction in another way. Some of the buyers might cheat by forming a bidding ring.(14) Suppose for example, that Agents K, K + 1, . . . N get together before the public auction and agree not to bid against one another. Instead they hold a private auction among themselves to determine who is willing to pay the most and agree that he will be the one to buy. Since Agents K, . . . N − 1 do not bid above p_{K-1}, Agent N will be able to buy the Old Master for about p_{K-1}, so that her share of the surplus from the sale is $p_N - p_{K-1}$ rather than $p_N - p_{N-1}$. Agent N will then compensate Agents K through N − 1 with side payments, so that all in the bidding ring are better off, although, of course, the seller is worse off.

The bidding ring is a cooperative coalition, but it is also a deviation from the cooperative coalition of the auction itself, in that the bidding ring does not follow the auction rule of non-cooperative bidding. It is also a coalition designed mainly to extract benefits at the expense of others, namely, in this case, at the expense of the seller. For these reasons, a bidding ring would probably be considered an undesirable and antisocial instance of cooperation. It might even be illegal, perhaps regarded as fraud.

The auction illustrates the mixture of cooperation and competition that we find throughout a modern economy. Of course, auctions play an important role in the economy. Many financial prices, such as those of common stocks, are largely determined by auctions. The Northwest Louisiana cattle farm I grew up on sold many of its feeder cattle at auctions. I remember being taken to the auction when "my" calf – a gift from the farm that I had been expected to feed – was sold at auction. The money went into my college fund. I was about eight.

But the example of the auction is important in another way. It is an example of the determination of an efficient market price. Markets and market prices are pervasive in modern economies. Leon Walras,(15) one of the founders of modern market theory, argued that, if there is sufficient competition, market prices are determined *as if* the market system were one great auction. For each type of good there is a demand curve, determined by the reservation prices of buyers much along the lines of Malthus's willingness to pay listing or the marginal utility schedule in the exchange theory of Jevons and Menger. For each type of good there are many potential sellers, each with her own reservation price, and those reservation prices together are a supply curve. The price that is realized is both the willingness to pay of the marginal buyer whose willingness to pay is least among those who do buy, and at the same time the reservation price of the marginal supplier, whose reservation price is the highest among all of those who do sell something.

This idea, that a market system determines prices as an auction does, is more than an analogy. Much of microeconomic theory is devoted to extending the results we saw for the auction to competitive markets in general: allowing for production and the sale of multiple units, exploring the relation of the cost of production to the reservation prices of suppliers, exploring the circumstances under which market prices are and are not efficient. In this study conspiracies in restraint of competition, such as the bidding ring, are at best an imperfection, and the study of the implications of such imperfections is also an important part of microeconomics. But it is no less true (though this point is often missed in the literature of economics) that the market system is a system *designed* to organize cooperation on a larger scale, relying on non-cooperative action, within the rules of the game, to realize the objectives of the cooperators. We will return to this idea in Chapter 8.

The discussion in this section is somewhat speculative in one sense. We have argued that the participants in the auction, as a cooperative coalition, rely on a rule of non-cooperative bidding because it requires relatively little information. This is a familiar theme in non-cooperative game theory and in the economics of auctions, but the introduction of limited information into cooperative game theory is a rather new frontier of research. Accordingly, we will return to the example in a future chapter.

CHAPTER SUMMARY

We have seen in this chapter that exchange can occur because it makes both of the parties to the exchange better off. Some of the controversy along the way has been about the interpretation of the term "better off." At the founding of economics, it seemed that utility or usefulness had very little to do with exchange – that value in exchange was quite unrelated to value in use. This paradox was resolved a century later by a more careful analysis of utility in its relation to exchange. We learned that the value in exchange of a good or service should be related to its marginal utility, not to its total or average utility. This also provided a new insight on the importance of exchange. Smith had understood that exchange was useful because it allowed more efficient production. The marginal utility approach showed that, even with given endowments (that is, after production was finished), exchange could lead to mutual advantage. This idea, in the work of Jevons and Menger, was based on a concept of subjective benefit or utility.

In the light of game theory, though, we realize that exchange is a cooperative phenomenon – in a world of purely non-cooperative action, there could never be any exchange. And yet exchange is a cooperative phenomenon that can often arise from non-cooperative commitments, such as the bidding in an auction. The exchange ultimately made is cooperative, and so is the organization of the auction itself; yet the auction is conducted by way of non-cooperative bidding in order to realize these cooperative goals. This provides us with a suggestive hint about the reasons behind the mixture of cooperative and non-cooperative activity that we see in the actual world, and we will return to it in later chapters.

One thing that we have not yet allowed sufficiently for is the role of uncertainty and information, and that will be the topic of the next two chapters.

SOURCES AND READING

(1) Neale's essay was published as Walter C. Neale (1957), "Reciprocity and redistribution in the Indian village: Sequel to some notable discussions," in Karl Polanyi, Conrad M. Arensberg and Harry W. Pearson (eds), *Trade and Market in the Early Empires*, New York: The Free Press.

(2) Adam Smith's remarks on the propensity to exchange are to be found in op. cit., Chapter 1 of this volume, note 7. (3) Malthus's passage is from a little remembered writing, Thomas Malthus (1800), "An investigation of the cause of the present high price of provisions," London: Davis, Taylor and Wilks. It may be read online at the archive maintained by McMaster University in Canada, available at http://socserv.mcmaster.ca/econ/ugcm/3ll3/malthus/highpric.txt, as of 8 November 2013.

(4) Modern utilitarian thought stems from the works of Jeremy Bentham. Probably the key reading is Jeremy Bentham (1789), "An introduction to the principles of morals & legislation," Oxford: Clarendon Press, available from the Library of Economics and Liberty at http://www.econlib.org/library/Bentham/bnthPML.html, as of 19 May 2013. (5) The Mills – James and John Stuart – are also important especially for the influence of utilitarianism on economics. John Stuart Mill is sometimes called the second founder of utilitarianism; see John Stuart Mill (1863), *Utilitarianism*, Kitchener, ON: Batoche Books. (6) Jevons's approach was summarized in a remarkable research paper, William Stanley Jevons (1866), "Brief account of a general mathematical theory of political economy," *Journal of the Royal Statistical Society*, **XXIX** (June), 82–7. Read first in Section F of the British Association, 1862, the paper had little influence for a decade afterwards. This may also be read online at the McMaster University archive at http://socserv.mcmaster.ca/econ/ugcm/3ll3/jevons/mathem.txt, but the version at the Marxists.org Political Economy Archive, at http://www.marxists.org/reference/subject/economics/jevons/mathem.htm, may be a little easier to read; both were last available 8 November 2013.

(7) Menger, op. cit., Chapter 2 of this volume, note 6. (8) For Edgeworth's contribution see especially Francis Edgeworth (1881), *Mathematical Psychics*, London: C. Kegan Paul. (9) Paul A. Samuelson (1948), "Consumption theory in terms of revealed preferences," *Economica*, **15**(60), 64–74. (10) Smith's words are found in op. cit., Chapter 1 of this volume, note 7. (11) The approach to market theory via the theory of cooperative games originated with Martin Shubik (1959), "Edgeworth market games," in Albert W. Tucker and Robert D. Luce (eds), *Contributions to the Theory of Games, Volume IV*, *Annals of Mathematics Studies, No. 40*, Princeton, NJ: Princeton University Press, pp. 267–78. It was extended by a number of studies in the 1960s and 1970s.

(12) I experienced an auction conducted like the one in Staten Island, New York, in 1974. It turns out that this auction is approximately equivalent to a second-price sealed bid auction, studied by Vickrey in an early contribution to the economics and game theory of auctions, and considered an ideal form for an auction such as this one. The reason that the Staten Island auctioneer chose that form seems to have had more to

do with speed, however, as there were many items to sell, and an auction conducted in this way goes a bit faster than a conventional "English auction" in which the buyers call out prices that they are willing to pay. Auctions were intensely studied in economics and game theory in the late twentieth century and since. A key founding paper in this literature was William Vickrey (1961), "Counterspeculation, auctions, and competitive sealed tenders," *Journal of Finance*, **16**(1) (March), 8–37. A good overview from the end of the twentieth century is John McMillan, Michael Rothschild and Robert Wilson (1997), "Introduction" (to Special Issue on Market Design and Spectrum Auctions), *Journal of Economics and Management Strategy*, **6**(3) (Fall), 425–30, and the papers in that special issue. (13) For a less upbeat assessment see Paul Klemperer (2002), "How (not) to run auctions: The European 3G telecom auctions," *European Economic Review*, **46**(45) (Apr.), 829–45. (14) For an example of a bidding ring, see John Asker (2010), "A study of the internal organization of a bidding cartel," *American Economic Review*, **100**(3), 724–62. (15) Walras's ideas are to be found in Leon Walras (1874), *Elements of Pure Economics, Or, The Theory of Social Wealth*, translated in 1954 by William Jaffe, Homewood, IL: Irwin.

5. Further benefits of working together: sharing risk

In all that has been said so far, knowledge and decisions have played important roles. Decisions to exchange or to produce for oneself, work together or independently, to offer a higher or a lower price, all have to be made on the basis of our knowledge about the results of those decisions. Often, though, we must make our decisions before we know for certain what the result will be. That is, we have to cope with uncertainty. The ways that people cope with uncertainty affect the ways that people work together, and conversely. To further advance our understanding of how people work together, the next step will be to consider how they cope with uncertainty.

UNCERTAINTY, PROBABILITY, AND RISK

Human beings have lived with uncertainty as long as we have lived. The classical pagans saw the uncertain future as determined by the unpredictable gods, and personified uncertainty as a goddess, Fortuna. For those who were favored by Fortuna, life could be good and power and influence might come, but she was a fickle goddess who might change her mind at any time. In more modern times, we have developed mathematical tools that enable us to take uncertainty into account as we consider the future and make decisions. These mathematical tools center on the concept of probability.

The mathematics of probability has been developed in the past few centuries. Some important early ideas came from the philosopher-mathematician Blaise Pascal, who used the ideas (among other things) to argue that it is wise to believe in the existence of God and in an afterlife even if those matters are uncertain. A number of important mathematicians, including Fermat, Bernoulli, and LaPlace made important contributions, and an economist must also mention William Gosset, pen name "Student," who advanced our understanding of statistics while advising the Guinness brewery on quality control for beer!

Probability is a numerical measure of the likelihood of an uncertain event. Whenever we consider an event that can have more than one outcome, such as throwing a pair of dice or conducting an experiment in medical research, we may judge that one outcome is more likely than another. Anyone who experiments with a pair of dice will soon discover that it is much more likely that the number 7 will be seen as the total number of spots when the dice are thrown than it is that the number 3 will be seen. But how much more likely? In particular cases, it may be easy to give this judgment a numerical value as a probability. Visualize a red die and a green die. When the dice are thrown, 36 distinct outcomes may be seen: the red die may show a 1, 2, . . . 6, and the same for the green die. There are just two ways that we may observe a 3 for the total of the two dice: the red shows 1 and the green shows 2 or vice versa. Thus, we say that the probability of a 3 is 2 in 36 or 2/36 = 1/18. We find that there are six outcomes in which the total is 7, so we say the probability of a 7 is 6/36 = 1/6.

Probabilities are always in the range from 0 to 1, and a more likely event will have a larger probability. If we have a list of all of the possible outcomes of a particular event or experiment, the total of the probabilities of those different outcomes must be 1. Thus, we immediately know that the probability that we will *not* get a total of 3 when we throw two dice is 17/18, and the probability that we will *not* get a 7 is 5/6.

If we were to experiment by throwing the dice 100 times, we would (most probably!) see that the proportion of throws for a 7 would be close to, though not exactly, 1/6; and that the proportion of throws for a 3 would be close to 1/18. The proportion of throws for a 3, out of a large number of throws, is the relative frequency of the 3, and the proportion of throws for a 7, out of a large number of throws, is the relative frequency of the 7. We see that probability is closely related to relative frequency. If we were to experiment by throwing the dice 1000 times, we would find that the relative frequencies would tend to be closer still to 1/18 and 1/6; with 10000 throws closer still. As the number of throws increases, the relative frequencies will tend to be closer and closer to the probabilities. In mathematical language, the relative frequency approaches the probability in the limit as the number of repetitions increases without bound. For experiments that can be repeated as many times as we may wish, as throwing dice can, we can confidently identify the probability of an outcome with the limit of its relative frequency. This provides the experimental evidence for many important estimates of probabilities, as, for example, in medical science.

For some uncertain events, however, evidence of this sort will not be available. A stockbroker once said to me that by 2030, automobiles would be propelled by electricity generated by an onboard fuel cell. Then again,

I recall a magazine article in the 1950s that said that nuclear fusion would be used to propel automobiles. Now, I don't suppose that either of those predictions is very likely, but it does seem to me that fuel cells are more likely to be used than nuclear fusion. How much more likely? I would like to have the probabilities in the two cases, but this is not an experiment that can be repeated a large number of times! One way to deal with uncertainty like this is to make the best guess one can – that is, to make an estimate of the probability based on whatever one may know that might be relevant. This is a subjective probability. For what it is worth, my subjective probability for the prediction that nuclear fusion will propel cars by 2030 is 0, and for fuel cells it is 2/1000 of 1 percent. I didn't buy the stock the broker wanted to sell me, either.

Subjective probabilities are somewhat controversial. Well-informed people will sometimes make quite different subjective estimates of the probability of the same event. Some scholars suggest that for some events, there simply is no reasonable way to determine the probabilities of the outcomes. Perhaps my example of cars propelled by fuel cells would be an example of that. The American economist Frank Knight is associated with this view, and uncertainty that cannot be expressed in terms of probabilities is often called "Knightian uncertainty." Against this, others, such as Leonard Savage, have argued both that it is always possible to make a rational estimate of the probability of an uncertain event and that this is *necessary* for rational decision-making. Perhaps few human beings are rational enough always to follow Savage's advice, but on the other hand, it is not at all clear what the practical implications of Knightian uncertainty might be.(1) Accordingly, in the balance of this book, we will take the easy way out and always express uncertainty in terms of probability. In economics, the common word risk refers to uncertainty expressed in terms of probability.

We will not need to dig deeply into the mathematics of probability, but there are a few related concepts that will be useful to us. Here again we will proceed by example. As a first step we consider how to make rational decisions when we face uncertainty.

MANAGING RISK

Probabilities are used as a tool for understanding events and ideas that are uncertain. We may use them simply as a tool of understanding, for example in scientific research, where we are only concerned to discover the truth. But they are also useful if we need to make decisions when the benefits and costs of the decisions are uncertain. An uncertain event that has benefits and costs for me is called a prospect. Suppose, for example, that someone offers me a wager: I will throw a pair of dice and he will pay me $10 if the dice show a total of three spots; and otherwise I lose $1. Should I take the bet? In order to decide, I will compute the mathematical expectation, also called the expected value of the prospect. The expected value is the weighted average of the payoffs, where the weights are the probabilities of the different payoffs. We recall that the probability of a 3, when two dice are thrown, is 1/18. Thus, the probability that I will win the bet is 1/18, and the probability that I will lose is 17/18. The expected value of my winnings is $(10)(1/18) - (1)(17/18) = (10 - 17)/18 = -7/18$ – a loss of slightly less than half a dollar. So I will refuse the bet. Suppose instead that the person offers me the $10 payoff if a 7 shows, and I lose $1 if the dice show any other total. The probability of a 7 is 1/6, and the probability of something other than a 7 is 5/6, so that the expected value is $(10)(1/6) - (1)(5/6) = (10 - 1)/6 = \$1.50$, a win of $9/6 = \$1.5$ dollars. So I will take that bet.

The expected value of a prospect is an important tool for decision-making. We might think that people try to make decisions in ways that give them the largest expected value for the net benefits from a prospect – as in the example of betting that a 3 will be thrown, where "reject the bet" gives me a larger expected value for the first bet (0 rather than –7/18), or in the example of betting that a 7 will be thrown, where "accept the bet" gives a larger expected value than "reject the bet." But money payoffs may not be the whole story. After all, we want money for what it will buy, and we want what money will buy for the subjective good states of mind, the utility, it will bring us. The utility of the payoffs may not be proportionate to the money value of the payoffs. More concretely, if I take the bet that a 7 will be thrown, I do not walk away with the weighted average payoff of $1.50 – I walk away with either a gain of $10 or a loss of $1. I may prefer not to take the risk of losing my dollar even though the average payoff, if I take that risk, is $1.50. In utility terms, it may be that the utility of winning $10 (weighted by the probability of 1/6) is not enough to compensate me for the utility I give up if I lose $1 (weighted by the probability of 5/6). In such a case I am said to be risk-averse. It is believed that risk aversion plays a role in many business and other decisions. A decision-maker who always chooses the option with the higher money expected value is said to be risk-neutral. In what follows, for simplicity, we will assume that decision-makers are risk-neutral unless it is stated otherwise.

We next consider an example to illustrate how two risks may or may not be related.

INDEPENDENCE AND CORRELATION

For our example, Ruby and Pearl are homeowners on Dinky Lane. Dinky Lane is a wooded neighborhood, and there is always some danger that a tree will fall and damage a house if the weather is bad. This is an uncertain event – the tree may or may not fall and the houses may or may not be damaged. We consider four different cases:

Case 1. Their houses are quite close together and there is a single tree that might fall and damage them. If it falls it will damage both houses. The probability that it will fall this year is 0.02, and so the probabilities of damage to the houses are as shown in Table 5.1. Thus, if we knew that Pearl's house has been damaged then we can infer at once that Ruby's house has also been damaged, and conversely. In such a case we say that the damage events are correlated, and in this instance the correlation is perfect, expressed by a correlation coefficient of 1.

Table 5.1 Probability of damage in Case 1

Both houses	0.02
Ruby's house	0.02
Pearl's house	0.02
Neither house	0.98

Case 2. Again, the houses are close together and a single tree threatens them both. In this case, however, they are on opposite sides of the tree, so that if the tree falls northward it will damage only Ruby's house, and if it falls southward it will damage only Pearl's. There are no other dangers of damage to either house. The probability that the tree falls northward is 0.01, the probability that it falls southward is 0.01, and the probability that it does not fall is 0.98. The probabilities of damage to the houses are as shown in Table 5.2.

Table 5.2 Probability of damage in Case 2

Both houses	0
Ruby's house	0.01
Pearl's house	0.01
Neither house	0.98

If we learn that Ruby's house has been damaged, then we can infer that Pearl's has not, and conversely. This is an instance of a negative or inverse correlation, and again it is a perfect negative correlation. It would be expressed by a correlation coefficient of –1.

Case 3. Again, the houses are close together and are threatened by a single tree. In this case the tree can fall north, south, east or west. If it falls northward it will damage only Ruby's house, if southward only Pearl's, but if it falls eastward or westward it will damage both houses. The probability of a fall in any of the four directions is 0.01 and the probability that the tree will not fall at all is 0.96. The damage probabilities are shown in Table 5.3.

Table 5.3 Probability of damage in Case 3

Both houses	0.02
Ruby's house	0.03
Pearl's house	0.03
Neither house	0.96

If we learn that Ruby's house has been damaged, we know that the tree has fallen in one of three directions, north, east or west, so we can infer that it is quite likely that Pearl's house has also been damaged, although we are not quite sure. This is a case of imperfect, positive correlation and would be expressed by a correlation coefficient between 0 and 1. However, we will not compute the correlation coefficient for this case, since our purpose is really to focus on the fourth case.

Case 4. In this case the houses are too distant from one another to be damaged by a single tree, but there is a tree in each yard that may fall and damage the house with a probability of 0.02. Both, neither, or just one of the trees may fall. In this case, the probability that both will fall is the product of the probabilities for each tree separately, $(0.02)(0.02) = 0.0004$; and the probability that exactly one tree will fall (and not the other) is $0.02 + 0.02 - (0.02)(0.02) = 0.0396$.[(2)] The damage probabilities are as shown in Table 5.4.

In this case, we say that the risks that the trees will fall are independent risks. They are not correlated and their independence could be expressed by a correlation coefficient of 0. If we are informed that Ruby's house has been damaged, we cannot infer anything about the probability of damage to Pearl's house that we did not already know, namely that the probability is 0.02.

At first, independent risks may seem more complicated, but in realistic

Table 5.4 Probability of damage in Case 4

Both houses	0.0004
Ruby's house	0.02
Pearl's house	0.02
Neither house	0.96

examples independent risks turn out to be simpler in a mathematical sense. They are also quite common. In the first three examples we assumed that the houses were damaged by the same tree, but in a world of many houses, many lanes and many trees, that would be unusual. More generally, in a big world in which many people and many causal processes act independently, independent risks are common, and even if some risks are not quite independent they may be approximately independent to a close approximation. But what can Ruby and Pearl do about their risky situation?

INSURANCE: SHARING RISK

Expected values, correlated and independent risks are tools for decision-making when prospects are risky. So far we have considered only individual decisions. In some cases, however, people may be able to reduce the risk they face by working together interdependently. Perhaps surprisingly, when risks are independent, the people are most profitably interdependent. To illustrate this, we return to Ruby and Pearl and the risks they face of damage to their homes. We will assume the risks are independent, as in Case 4 in the section above, and to keep things simple, we assume that Ruby and Pearl are both risk-neutral – they will choose among their alternatives the one alternative that has the greatest expected value of money payoffs.

As before, we assume that each faces a risk of damage to her home from bad weather. We suppose that the damage will cost $5000 if repaired immediately, but $50 000 if the repair is postponed, since leakage and settling will cause further damage. No insurance policies are available, since Pearl and Ruby live in a remote country in which insurance has never been invented. Thus, in order to be prepared for damage, each householder must have $5000 available to pay for immediate repairs. They both have the money, but they would prefer to invest it in assets that would tie up the money in such a way that it would not be available to do immediate repairs, but which pay a better rate of return. The bank offers each of them a choice of two assets: a demand deposit, which pays a rate of return of 2 percent but can be withdrawn to pay for immediate repairs if damage occurs, and a saving certificate. The certificate pays 10 percent interest but, if damage occurs, repairs will have to be postponed until the certificate is mature so that the damage of $50 000 will occur. Thus, each of the two homeowners is (separately) making a decision with an uncertain, risky outcome. We can treat this as a game, with nature as the "opponent," understanding that nature always plays her strategies with specific probabilities. This is called a game against nature. Their game against nature is shown in Table 5.5. The homeowner's strategies are to place the $5000 of funds in the demand deposit or the certificate. Nature's strategies are to cause damage to the homeowner's house or not, with probabilities 0.02 and 0.98. The payoffs are the homeowner's financial assets following both decisions, after receipt of interest and payment for repairs if any. We see that the homeowner's best response strategy is to choose the demand deposit. They are keeping their own funds available to cover the cost of the improbable bad event. This is called self-insurance.

Table 5.5 A Game Against Nature

		Nature		
		Good weather	Bad weather	Expected value
Homeowner	Demand deposit	5100	0	4998
	Certificate	5500	−44 500	4500
	Probability	0.98	0.02	

Now, suppose that Ruby and Pearl get together and make a cooperative agreement to share the risk. That is, they agree that if one of them experiences damage, they will share the $5000 cost of immediate repair equally. If both houses are damaged, the agreement is off. Another way to accomplish this would be a bet: Ruby says "I'll bet you $2500 that my house is damaged and yours is not." Either way, the bet or the agreement would have to be enforced, since it is a cooperative arrangement (unless the two homeowners could trust one another to honor the agreement, or the bet, without enforcement). The agreement could take the form of a contract that would be enforceable in a court of law.

This cooperative arrangement results in a new investment game for each of the homeowners. On the one hand the strategy of investing the $5000 all in certificates is ruled out, since the homeowner would then be unable to share the payment for immediate repair as the agreement requires.

However, there would be a new strategy: investing half in certificates and half in demand deposits. If both experience damage, neither will have enough liquid assets to do the immediate repairs, and the greater cost will have to be paid, but since the risks are independent, the probability of this is very small. The new game against nature is shown as Table 5.6.

Table 5.6 A Game Against Nature for cooperative coalition

		Nature			
		No damage	Damage to one	Damage to two	Expected value
Homeowner	Demand deposit	5100	2550	0	4998
	Half certificates	5300	2750	−44700	5179
	Probability	0.96	0.0396	0.0004	

We see that the homeowners' best response is now to place half of their $5000 funds in the demand deposits and half in the certificates. This yields an expected value of $5179, which is better than the expected value of $4998 they could obtain without the cooperative arrangement. In effect, Ruby and Pearl have discovered insurance. Their cooperative agreement is what is called mutual insurance. In the real world there would be many more than two people in the agreement, and the agreement would apply to a wide variety of risks – any risks of damage from bad weather and such other causes as fire, for example. And this is good, because the benefits of mutual insurance increase with the number of different risks shared, so long as they are independent risks.

In real economic history, insurance has often been pioneered by mutual organizations among the insured, but another possibility is that a profit-seeking company might form individual contracts with Ruby and Pearl (and a large number of other people) to provide them with insurance for a price they would find satisfactory. The profit would arise from the increase in benefits as larger and larger numbers of people share their independent risks.

The total payoff to both Ruby and Pearl with the mutual insurance coalition thus is $10358, while with self-insurance and no coalition the total payoff is $9956, so the coalition creates value of $402. It seems natural to assume that this will be equally divided between the two of them, as $201 each, but we cannot be sure of that. It will depend on bargaining between the two of them. If they have equal bargaining power then they will divide the value created equally; otherwise perhaps not. (From this point on we will assume equal bargaining power, for simplicity.)

Notice, also, that the mutual insurance agreement between Ruby and Pearl enables them to place half of their funds in certificates, which are more profitable to them. This means that the funds are available to finance activities like operating a business and building houses. These activities tie up the funds for a relatively long time, so they cannot be financed as conveniently by demand deposits. In other words, sharing of independent risks increases the efficiency of the allocation of capital, that is, it allows the funds from saving to be put to more productive uses.

However, it is crucial that the risks are (very nearly!) independent. Suppose instead that they were perfectly correlated, as in Case 1 above. Then the game against nature with the cooperative agreement would be as shown in Table 5.7. The best response is to self-insure, with the payoff just as before.

Table 5.7 Game Against Nature with correlated damage

		Nature			
		No damage	Damage to one	Damage to two	Expected value
Homeowner	Demand deposit	5100	2550	0	4998
	Half certificate	5300	2750	−44700	4300
	Probability	0.98	0	0.02	

If the risks are perfectly correlated, then insurance does not create any value-added. Risks that are highly correlated in this way are often called systematic risks. Many economists believe that the economic crisis of 2008–09 occurred because, or was made worse because, the risks of many financial instruments (such as mortgage-backed securities) were incorrectly estimated, in that the risks were assumed to be independent and systematic risks were neglected.

In this section we have seen an example of yet another way that people can benefit from working together: by forming cooperative agreements to share independent risks.

PRICES, MARKETS, AND RISK

If Pearl and Ruby form a mutual insurance organization, they will be shareholders, with their own funds at risk. The mutual insurance organization would deposit its $5000 in demand deposits and, if neither home is damaged, it would earn the 2 percent rate of return and return a dividend of 2 percent to Ruby and Pearl along with their original investments. Many early insurance organizations were set up in this way. By contrast, a profit-seeking insurance firm would put its own capital at risk, charging Ruby and Pearl a fee for insurance services. The fee would have to be at least enough to compensate the insurance firm for giving up the higher return on certificates* (or other assets) in order to maintain enough readily available funds to pay claims, along with other costs. Then Ruby and Pearl would be free to use their own funds as they might choose, net of the fees they would pay. As we recall from Chapter 3, the fee charged by the insurance company would be limited by competitive alternatives. The formation of a mutual insurance company is one of the competitive alternatives available to Ruby and Pearl. If the insurance company fee is enough to reduce the total benefits to Ruby and Pearl below $402, they could form a mutual insurance organization, and not deal with the insurance company. In the real world, other companies will often be the more important competitive alternatives for Ruby and Pearl.

In the modern world, contracts of insurance are available in the market at prices that reflect their costs and benefits, and will be higher when the event that is insured is riskier. Prices in many other markets are influenced by risk. For example, when a banker makes a loan, the interest demanded by the banker will reflect the risk that the loan will not be paid back, and also the risk that inflation will reduce the value of the monetary units in which it will be paid. This in turn will influence many other prices of goods and services. For example, the cost of house construction, which often relies strongly on borrowed money, will reflect the interest rate on fairly risky loans.

However, markets for insurance have two well-known complications, which may sometimes be observed in other markets in which prices are influenced by risk. They are called moral hazard and self-selection.(3)

Moral Hazard

Moral hazard refers to the tendency for people to make different decisions when they are insured against risk than they would make if they were not insured. In particular, they may decide not to take precautions that would reduce the risk, precautions that they otherwise would take. We can illustrate this in the example of Ruby and Pearl. Suppose that Ruby

* In fact, insurance organizations are important investors, and managing their investment portfolios so as to be able to pay claims while earning the best return on their reserve funds that is consistent with that objective is a major task for any insurance organization.

can have her tree stabilized by cables, reducing the risk of damage, for $40. This will reduce the probability of damage to her house to 1 percent. If there is no insurance, this means that self-insurance will leave her with an expected value payoff of (0.99)(5100) + (0.01)(0) – 40 = 5009, so it will make sense for Ruby to take this precaution. If there is a mutual insurance coalition, Ruby's precaution will increase the value of the coalition to $10428, a value-added of $70. However, with equal bargaining power, Ruby gets only half of that, $35, which is less than the cost of stabilizing the tree, so Ruby will not do it. Thus, the insurance policy causes Ruby to change her decisions in a way that increases the total risk. This is moral hazard.

Suppose both homeowners have their trees stabilized. This will reduce the probabilities of damage still further – the probabilities in the bottom line of Table 5.6 would be 0.98, 0.0199, 0.0001. Without reconstructing the table yet again, this would increase the value of the coalition to 10408 = 2*5204. Thus, assuming that that Ruby and Pearl decide non-cooperatively whether or not to stabilize their tree, then we have the non-cooperative game shown in Table 5.8.

Table 5.8 A non-cooperative game of precaution

		Pearl	
		Stabilize	Don't
Ruby	Stabilize	5204,5204	5174,5214
	Don't	5214,5174	5179,5179

This game is a social dilemma, like Prisoner's Dilemma or the Taking Game from Chapter 3, in that Ruby and Pearl each have dominant strategies, and when they play the dominant strategy the Nash equilibrium that results is different from the cooperative solution. But they are both better off if they both choose the dominated strategy of taking the precaution and stabilizing their trees. Indeed, moral hazard is usually thought of as a Prisoner's Dilemma-like game in which the dominant strategy is for nobody to take the precaution and the cooperative solution is for all to take the precaution. One way to deal with the problem might be to adopt a more complex insurance contract, one that might require the homeowners to take the precaution or provide an insurance payment that would share the cost of the precaution.

In the perspective of this chapter, moral hazard may seem rather odd, in that (we assume) the homeowners have already arrived at a cooperative agreement but now choose non-cooperatively not to install the precaution.

We will return to this puzzle in Chapter 7. For now, however, we may note that insurance contracts often do include rules requiring the insured person to take certain precautions, and other rules that reward the insured person with a cheaper rate if precautions are taken, or that pay all or part of the cost of the precautions, in order to reduce the risk shared by all who are insured. An example of this is the fact that many medical insurance policies repay not only the cost of (unpredictable) catastrophic illness, but also the cost of (predictable) routine medical care – which is, among other things, a precaution against catastrophic illness. But medical insurance also provides a key example of self-selection.

Self-selection

Self-selection is a problem that can occur in many insurance markets but is especially a problem in health insurance. It arises because people know more about their own health risks (and their own risks generally) than the insurer does. As usual we will explore the point with an example. For the example, Tom, Dick, and Harry all face health risks that may cost them $100 000 for treatment. They are considering purchasing health insurance policies to pay for the treatment. For Tom, the probability of the health crisis is 0.01, for Dick, 0.02, and for Harry, 0.03. These probabilities are independent. Tom, Dick, and Harry know their probabilities but the insurance company does not. The insurance company knows only that one each of the clients has a probability of 0.01, 0.02, and 0.03, but doesn't know which. If they all buy insurance, the insurance company faces risks of paying $100 000, $200 000, or $300 000, or nothing. Table 5.9 shows the probabilities and the expected value for the insurance company.(4) Although it does not know *which* of the three men has probabilities 0.1, 0.2, and 0.3, the insurance company knows that the probabilities are distributed in that way and so can compute the overall probabilities that one, two, or three claims will be made.

This is an example of asymmetric information. Tom, Dick, and Harry each know something that the insurance company does not. In general, when different individuals have different knowledge and are uncertain about different things, then by definition we have asymmetric information. When some have information that would affect the decisions of others – as in this example – asymmetric information can be a problem for people trying to work together.

This insurance coalition provides Tom with a gross benefit (expected value of risk avoided) of $1000; Dick a gross benefit of $2000, and Harry a gross benefit of $3000, for a total of $6000, at a cost (expected value of payouts) of $5898.64. Thus, the coalition generates added value of

Table 5.9 Health risks for insuring Tom, Dick, and Harry

Who is Ill	Probability	Payoff	Probability × Payoff
None	0.941094	$0.00	$0.00
All three	0.000006	−$300 000.00	−$1.80
Just one, i.e., just Tom *or* Dick *or* Harry	0.057	−$100 000.00	−$5680.51
Just two, i.e., Tom *and* Dick, Tom *and* Harry, *or* Dick *and* Harry	0.001081653	−$200 000.00	−$216.33
Expected value			−$5898.64
Per person			−$1966.21

$101.36 = $6000 − $5898.64. This modest value-added reflects the fact that only three people are being insured. Here as always, sharing a larger number of independent risks results in a greater value-added, and in a realistic example the insurance would share the risks of many thousands.

In order to cover its expected costs, the insurance company must charge the three people $1966.21 each for their health insurance. But now Tom will have to decide whether it is worthwhile for him to purchase insurance at this price. In effect, Tom will play a game against nature that is shown as Table 5.10. We see that Tom's best response (if he is risk-neutral) is not to insure. Thus, Tom will select himself out of the risk-sharing group by declining to buy insurance.

Table 5.10 Tom's Game Against Nature

Strategy		Nature		
		Ill	Well	Expected value
Tom	Don't insure	−$100 000.00	0	−$1000.00
	Insure	−$1966.21	−$1966.21	−$1966.21
	Probability	0.01	0.99	

If Tom does not purchase insurance, the insurance company may still try to sell policies to Dick and Harry. The fact that Tom is the one who dropped out makes it clear that Tom has the low probability of 0.01, but the insurance company still does not know which of the other two has the higher probability. To insure just Dick and Harry, their risks will be as shown in Table 5.11. From this table we see that, in order to cover the cost of insuring Dick and Harry, the health insurer will have to charge each of them $2500.00. This is more than the charge would have been to insure all three, which was $1966.21, because Tom, the healthiest of the three, is no longer in the pool. A health insurance pool that is healthier, on the average, can be insured at lower cost. Here the cost rises because the pool is less healthy, on the average.

Table 5.11 Risks of insuring Dick and Harry

Who is Ill	Probability	Payoff	Probability × Payoff
None	0.9506	0	0
Dick and Harry	0.0006	−$20 000.00	−$120.00
Just one, i.e., Dick *or* Harry	0.0488	−$100 00.00	−$4880.00
Expected value			−$5000.00
Per person			−$2500.00

Now Dick must play a game against nature to determine whether to buy insurance at this price or not. He gains an expected value of $2000 = 0.02 × $100 000 from insuring, so insurance would have been the best response for Dick at the old price of $1966.21. At the new price of $2500, however, Dick's best response is to drop out of the coalition and not buy any insurance. That leaves only Harry, the least healthy of the three, in the market for insurance – and since there is no risk-sharing, there is no benefit from insurance for a single individual. In this example, at a uniform price that just covers the insurance company's costs for paying claims, *everyone* eventually selects himself out of the market for health insurance, and no health insurance exists. At each stage, the healthiest find the cost of health insurance more than its value to them, and so decline it; but this raises the cost for the rest and this leads others, less healthy but still healthier than the average insured person, to drop out in turn, so that in the end there is no health insurance. Of course, this is an extreme example. In more realistic models some health insurance is provided, but less than would be efficient.

But we know that a coalition for health insurance is efficient in this example – it generates a net value of just over $100. How is it that this value-creating coalition cannot be formed? Suppose that the three individuals were offered insurance policies at different prices: $975 to Tom, $1975 to Dick, and $2975 to Harry. Then it would be worthwhile for each of them to purchase the insurance – insurance would make each of them better off by $25 – and the insurance company would earn a profit of $26.36. To do that, however, the insurance company would have to know which probability applies to which client – that the probability of a health crisis to Tom is 0.01, to Dick 0.02, and to Harry 0.03. We have assumed that the insurance company doesn't have that information. Of course, insurance companies do attempt to find out what they can about the health risks of the people they insure. But this information is not free. (We will return to this point in more detail in the next chapter.) It is costly, so that the company is not likely to obtain enough information to eliminate the tendencies toward moral hazard and self-selection from insurance markets.

BANKRUPTCY

Uncertainty can also help us to understand one of the more puzzling institutions of modern economies: bankruptcy. Why indeed should anybody be excused from paying his or her debts? In some earlier societies, people who borrowed money and could not repay it might be enslaved or put in prison. But slavery is no longer legal and debtors' prisons have been abandoned. In their place we have the institution of bankruptcy. Why?

As usual we will proceed by example. John Doe and James Roe are interested in forming a coalition to produce and sell widgets. To produce widgets, it will be necessary to assemble higher order goods, such as machines and raw materials and such. The cost of these higher order goods is $100 000. John Doe has $50 000 for the purpose, and James has nothing. Thus, in order to get started, they will have to bring a third member into their coalition: a bank or other moneylender.

There is also some uncertainty about the conditions for making and selling widgets. The total value of the coalition is thus unknown. We suppose that the coalition's revenue from selling widgets – which is the gross value of the coalition – is $180 000 if conditions are good, with probability 90 percent; $100 000 if conditions are bad, with probability 5 percent; and $50 000 if conditions are very bad, with probability 5 percent. The expected value of the coalition then is $169 500. We suppose also that if James Roe does not join the coalition, he can operate as an individual and earn $65 000 (perhaps by working as a self-employed craftsman).

Production of widgets thus will take place in three stages. At the first stage John Doe and the bank make a preliminary agreement about a loan from the bank to John Doe and they commit $50 000 each, with which John Doe assembles the higher order goods needed for production. At the second stage, the conditions for production and sales of widgets, and thus the value of the widget-producing coalition, become known. At the third stage, depending on the conditions for widget production, James Roe may be added to the coalition and the value of the three-person coalition is distributed. Bargaining power is determined by the contracts John Doe signs with the bank and with James Roe, and these in turn are determined by the customary norms of the capitalist economy in which they live: the bank is granted its $50 000 back, plus an agreed rate of interest, and James Roe gets his competitive alternative, $65 000. We will suppose that the interest rate is 10 percent so that the bank is entitled by its contract to $55 000. As proprietor, John Doe gets whatever remains of the value of the coalition: he is the residual claimant. As such, he also bears the uncertainty of the value of the coalition, in most circumstances.

Now, suppose the conditions for widget production and sale are good, so that the value of the coalition is $180 000. James Roe gets $65 000, the bank gets $55 000, and John Doe, proprietor, gets $60 000, from which he can deduct his initial commitment of $50 000 for a profit of $10 000.

Next, consider the opposite extreme: conditions for widget production and sale are very bad. Then the value of the three-person widget-producing coalition is $50 000. Since he can do better individually, James Roe will not enter the coalition. There is no three-person widget-producing coalition at all in this case, and instead we see only a two-person coalition of the bank and the proprietor. The value of this coalition is 0. The higher order goods procured for widget production go to waste, and the payoff to the bank and to the proprietor are both zero – or, after deducting the $50 000 they have both committed at the beginning, the payoff is –$50 000 each.

The most interesting case is the one in which conditions are bad, but not very bad. In this case, the value of the widget producing coalition is $100 000. This creates a dilemma. If the bank insists on receiving the $55 000 to which it is entitled by the contract it has signed with John Doe, John will not have enough left over to pay James Roe the $65 000 that James can earn as an individual. James will not join in the three-person coalition to produce widgets, and, as in the previous paragraph, no widgets will be produced. Suppose, instead, that the bank agrees to settle for less than the $55 000 that its contract entitles it to. In particular, suppose the bank will accept $35 000 or less. Then James can be paid $65 000 and will join the coalition and widgets will be produced and sold for $100 000. Now, $35 000 is better than nothing, and nothing is what the bank will get if James cannot be hired for his competitive alternative of $65 000. Thus, it is in the bank's interest to relax the terms of its loan contract when conditions are bad but not very bad. However, once the contract between the bank and the proprietor is cancelled, bargaining power comes into the picture. There then is $35 000 to divide between the bank and the proprietor, and both may try to hold out for a bigger share of this diminished value. At this point, a bankruptcy court will settle the matter. As a rule, the proprietor will lose everything, and the company will become the property of the bank, under new management, so that the bank recovers what it can. Such a bankruptcy statute is a protection and benefit *to the lenders*, in that it minimizes the cost of lending that arises from the uncertainty of production and selling and of bargaining power in the event of default. In general, bankruptcy is efficient and is in the interest of the lender in cases in which the borrower can pay either the competitive wage to hire labor, or the payments due on its debt, but not both.

How will the bank do overall? It expects a repayment, with interest, of $55 000, with probability 0.9; of $35 000 with probability 0.05; and of nothing with probability 0.05. The expected value then is $51 250. This is an expected net rate of return of 2.5 percent on the bank's investment. Most of the 10 percent interest goes to protect the bank against bad conditions and consequent default by the proprietor. But notice that if there were no bankruptcy or any relaxation of the terms of the loan in the case of bad conditions, the bank's expected value would be $49 500, for a net loss. In that case, it would refuse to loan, or would require an even higher interest rate to loan.

How does the proprietor do overall? He expects the net profit of $60 000 with probability 0.9, and nothing with probability 0.1. This leaves an expected value of $54 000, from which we may deduct $50 000 for a net profit of $4000. This leaves the proprietor with an expected value rate of return of 8 percent. If he is risk-averse, this relatively high rate of return may nevertheless be sufficient to compensate him for taking the greater part of the risk of this enterprise. James Roe, as an employee "at will," does not join the coalition unless it is successful enough to pay him the $65 000 he would obtain otherwise, and so bears none of the risk of the enterprise, but also has no bargaining power and gets none of the net value the coalition generates. (Remember that this is a very simplified example!)

We have seen that bankruptcy plays a key part in an economy in which businesses may often be partly financed by borrowed money, and in which the success of the business is uncertain, which is to say, in any modern market economy. Bankruptcy is not a minor detail – it is a key institution of a modern market economy. But bankruptcy is not a perfect institution. In the example we have been considering, the net value of the widget enterprise to society is the sum of the expected values to the bank and to John Doe, that is, $5250.* But suppose that the proprietor must make an effort to prevent the gross value of the coalition in bad conditions from dropping from $100 000 to $90 000. This deterioration would reduce the bank's expected value payoff to $750 and would reduce the value of the enterprise to society to $4750. This would not influence John's decision, since his own expected value is not affected – he will be bankrupt if the conditions are bad in any case. Thus, borrowed money and bankruptcy "distort" the proprietor's decisions in ways that may be inefficient. But, as we have seen, the alternative can be worse. In this example, if there were no statute of bankruptcy and no bank lending, the widget production would not take place at all, since John and James do not have enough investment funds to get it started, and, as we have seen, there is a net social benefit from the formation of the widget-producing enterprise. We will have to trade off one sort of inefficiency against another.

Here, again, the bank will take some steps to reduce the uncertainty of its loan. It will investigate John's reliability and inspect his business plan. It may also obtain some information about the market for widgets and the technology of widget production, to better understand the business plan. However, here, again, information is not free, so the bank will not find it profitable to be perfectly informed – even if that were possible. Therefore, uncertainty will always be a part of any important business decision.

CHAPTER SUMMARY

To reiterate, uncertainty will always be an aspect of any important business decision. The mathematics of probability gives us tools to quantify uncertainty. When we can experiment many times with the same uncertain event, we can identify the probability of a particular outcome with the relative frequency of that outcome as the number of experiments becomes very large. Otherwise, we can only rely on subjective estimates of probability.

* This ignores any consumers' surplus or consumers' net benefit from buying widgets. This is a simplifying assumption. To make the example more realistic, we might include the widget buyers as members of the widget coalition. In that case, the social benefits of the enterprise would be even greater. We might also allow for more than one employee, for some flexibility to employ more or fewer employees, more or less higher order goods as substitutes for more employees, and the choice among techniques of production. All of these complications would make the example more realistic but would not change the point the example is meant to illustrate: that bankruptcy is a key institution for a market economy, when we allow realistically for uncertainty.

Uncertain events with costs and benefits are called prospects or risks. To express the overall value of an uncertain prospect, we may compute the expected value, that is, the weighted average of the benefits in different outcomes, weighted by the probabilities of the outcomes. However, people who are risk-averse may value a risky prospect less than its expected value, and there may be other people, risk lovers, who value a risky prospect more than its expected value. People who are neither risk-averse nor risk-lovers are said to be risk-neutral.

When risks are independent, people can benefit by agreements to share the risk. This is insurance. However, markets for insurance face two problems: moral hazard and self-selection. These difficulties arise because the insurance organization cannot have complete information about the decisions and risks known to the individuals who are insured. Similar difficulties can arise in any financial market, such as a market for loans. In a market for business loans, with uncertainty, bankruptcy plays a key role, even though it can distort some business decisions away from efficiency. In a world of uncertainty, some inefficiency seems unavoidable, as arrangements that limit inefficiency from one point of view often will create inefficiency from another. We can only trade off one inefficiency for another and hope for a satisfactory compromise.

SOURCES AND READING

Among the topics of the first three sections, probability, mathematical expectation, and independent risks and correlation will be covered in much more detail in any standard introductory statistics text. (1) Frank Knight first set out his ideas on risk and uncertainty in Frank Knight (1921), *Risk, Uncertainty and Profit*, Boston, MA: Houghton Mifflin, available online from the Library of Economics and Liberty, at http://www.econlib.org/library/Knight/knRUP.html, as of 2 August 2012. Aside from Knight, there were other economists who questioned the applicability of probability as usually understood with regard to business decisions, notably John Maynard Keynes (1973), in *A Treatise on Probability*, London: Macmillan, originally published in 1921, available online from the Gutenberg Project at http://www.gutenberg.org/files/32625/32625-pdf.pdf, also as of 2 August 2012. Another was G.L.S. Shackle, who seemed to have been influenced by Keynes's ideas. His 1949 book, *Expectation in Economics*, Cambridge: Cambridge University Press, is probably the most important statement of his ideas. Also influenced by Keynes's thinking, I once tried to find a middle ground between the two views: Roger A. McCain (1983), "Fuzzy confidence intervals," *International Journal of*

Fuzzy Sets and Systems, **10**, 281–90, and Roger A. McCain (1987), "Fuzzy confidence intervals in a theory of economic rationality," *International Journal of Fuzzy Sets and Systems*, **23**, 205–18. Leonard J. Savage set out his theory of subjective probability in his 1954 book, *The Foundations of Statistics*, New York: Wiley Publications in Statistics. (2) Examples of damage to the homes of Ruby and Pearl are shown in the appendix to the chapter and also online at http://goo.gl/xAxHYU.

(3) The term "moral hazard" is a traditional one in insurance and has been traced back to the eighteenth century. The concept of self-selection in markets with asymmetrical information is found in George Akerlof (1970), "The market for lemons: Qualitative uncertainty and the market mechanism," *Quarterly Journal of Economics*, **84**(3) (Aug.), 488–500. Akerlof did not apply it to insurance, however. Michael Rothschild and Joseph Stiglitz did make that connection in a paper presented at a conference on imperfect information at Princeton in the spring of 1973 but not published until 1976: Michael Rothschild and Joseph Stiglitz (1976) "Equilibrium in competitive insurance markets: An essay on the economics of imperfect information," *Quarterly Journal of Economics*, **90**(4) (Nov.), 629–49. This may or may not have been the first expression of the idea but was certainly an important and seminal one. Akerlof and Stiglitz were honored, along with A. Michael Spence, for their work on asymmetrical information by the Nobel Memorial Prize in 2001. (4) Calculating the probabilities of these claims is a little complex so we will save it for the appendix to the chapter and also online at http://goo.gl/xAxHYU.

APPENDIX

Part 1

Master tables for the examples of damage to the homes of Ruby and Pearl

Table 5A.1 Table of constants and probabilities Master spreadsheet for the examples of damage to the homes of Ruby and Pearl Constants

ppp	0.02	Independent probability of damage to one house
c1r	5000	Cost of immediate repair
c2r	50000	Cost of delayed repair
rod	0.02	Rate of return on demand deposit
roc	0.1	Rate of return on certificate of deposit
wealth	5000	Wealth to be invested by each
ppc	0.02	Probability of correlated damage
pqq	0.01	Probability of damage if precaution is taken
cost	40	Cost of precaution

Table 5A.2 Game Against Nature for one homeowner

		Nature		
		Good weather	Bad weather	Expected value
Homeowner	Demand deposit	5100	0	4998
	Certificate	5500	−44 500	4500
	Probability	0.98	0.02	

Table 5A.3 Game Against Nature for cooperative coalition

		Nature			
		No damage	Damage to one	Damage to two	Expected value
Homeowner	Demand deposit	5100	2550	0	4996.98
	Half certificate	5300	2750	−44700	5179.02
	Probability	0.96	0.0396	0.0004	1
Value of coalition		10 358.04			
Non-cooperative case		9956			
Value-added		402.04			
Net per person, assuming equal bargaining power		201.02			

Table 5A.4 Game against Nature for coalition with correlated damage

		Nature			
		No damage	Damage to one	Damage to two	Expected value
Homeowner	Demand deposit	5100	2550	0	4998
	Half certificate	5300	2750	−44700	4300
	Probability	0.98	0	0.02	1
Value of coalition		9996			
Non-cooperative case		9996			
Value-added		0			
Net per person, assuming equal bargaining power		0			

Master tables for the example of moral hazard, in which a tree can be stabilized as a precaution

Table 5A.5 Game Against Nature for individual homeowner, with precaution

		Nature		
		Good weather	Bad weather	Expected value
Homeowner	Demand deposit	5100	0	5009
	Certificate	5500	−44 500	4960
	Probability	0.99	0.01	

Table 5A.6 Game Against Nature for cooperative coalition, assuming both take precautions

		Nature			
		No damage	Damage to one	Damage to two	Expected value
Homeowner	Demand deposit	5100	2550	0	5008.745
	Half certificate	5300	2750	−44700	5204.255
	Probability	0.98	0.0199	0.0001	1
Value of coalition		10 408.51			
Non-cooperative case		9978			
Value-added		430.51			
Net per person, assuming equal bargaining power		215.255			

Table 5A.7 Game Against Nature for cooperative coalition, assuming only one takes precaution

		Nature			
		No damage	Damage to one	Damage to two	Expected value
Homeowner	Demand deposit	5100	2550	0	4982.99
	Half certificate	5300	2750	−44700	5174.01
	Probability	0.97	0.0298	0.0002	1
Value of coalition		10348.02			
Non-cooperative case		10162.255			
Value-added		185.765			
Net per person, assuming equal bargaining power		92.8825			
Expected value gross of cost of precaution					
Demand deposit		5022.99			
Half certificate		5214.01			

Table 5A.8 Non-cooperative game of taking precautions, assuming the coalition takes place and assuming equal bargaining power

		Pearl	
		Stabilize	Don't
Ruby	Stabilize	5204,5204	5174.01,5214.01
	Don't	5214.01,5174.01	5179.02,5179.02

Part 2

Master tables for calculating the probabilities for the claims of Tom, Dick, and Harry

Table 5A.9 Probabilities and payoffs to Tom, Dick, and Harry without insurance

ptom	0.01	qtom	0.990
pdick	0.02	qdick	0.980
pharry	0.03	qharry	0.970
Tom	−1000		
Dick	−2000		
Harry	−3000		

Table 5A.10 Risks of insuring all three

	Who is ill	Probability	Probability Number	Payoff	Probability × Payoff
1	None	(0.99)(0.98)(0.97)	0.941094	$0.00	$0.00
2	Just Tom	ptom*(1-pdick)+(1-pharry)	0.009506	−$100 000.00	
3	Just Dick	pdick*(1-ptom)(1-pharry)	0.019206	−$100 000.00	
4	Harry	pharry(1-ptom)(1-pdick)	0.029106	−$100 000.00	
5	Tom *and* Dick	ptom*pdick*(1-pharry)	0.000194	−$200 000.00	
6	Tom *and* Harry	ptom*pdick*(1-pharry)	0.000294	−$200 000.00	
7	Dick *and* Harry	pdick*pharry*(1-ptom)	0.000594	−$200 000.00	
8	All three	ptom*pdick*pharry	0.000006	−$300 000.00	$1.80
9	Just one, i.e. just Tom *or* Dick *or* Harry	Just one of lines 2, 3, 4	0.057	−$100 000.00	−$5 680.51
10	Just two, i.e., line 5, 6, or 7	Just one of lines 5, 6, 7	0.001081653	−$200 000.00	−$216.33
Expected value					−$5 898.64
Per person					−$1 966.21
Surplus					$101.36
So if the company could charge $975 to Tom, $1975 to Dick and $2975 to Harry, all would be better off and the company would make a profit of					$26.36

Table 5A.11 Tom's Game Against Nature

		Nature		
	Strategy	Ill	Well	Expected value
Tom	Don't insure	−$100 000.00	0	−$1 000.00
	Insure	−$1 966.21	−$1 966.21	−$1 966.21
	Probability	0.01	0.99	

Table 5A.12 Risks of insuring Dick and Harry

Who is ill	Probability	Probability Number	Payoff	Probability × Payoff
None	qdick*qharry	0.9506	0	0
Just Dick	pdick*(1-ptom)(1-pharry)	0.019206	−$100 000.00	
Harry	pharry(1-ptom)(1-pdick)	0.029106	−$100 000.00	
Dick *and* Harry	pdick*pharry	0.0006	−$200 000.00	−$120.00
Just one, i.e., Dick *or* Harry	0.02+0.03-(0.02)(0.03)	0.0488	−$100 000.00	−$4 880.00
Expected value				−$5 000.00
Per person				−$2 500.00

Table 5A.13 Dick's Game Against Nature

		Nature		
	Strategy	Ill	Well	Expected value
Dick	Don't insure	−$100 000.00	0	−$2 000.00
	Insure	−$2 500.00	−$2,500.00	−$2 500.00
	Probability	0.02	0.980	

PART II

Information is not free

6. Information is not free

As we have seen in earlier chapters, people can often achieve mutual benefits by forming what game theory calls coalitions to adopt a common course of action. If the objective of the coalition is to realize a complex division of labor (Chapter 2) or for risk-sharing (Chapter 5), then a large number of individuals may be involved. In any coalition, transaction, or mutual agreement, the objective is to increase the value created so far as possible; but some information will be required in order to do so. Information is not free. Information is a higher order good,* and like other higher order goods, information must be produced by means of costly resources. There may also be special obstacles for the production of information as well. In the words of Defense Secretary Donald Rumsfeld, "there are known 'knowns.' There are things we know that we know. There are known unknowns. That is to say there are things that we now know we don't know. But there are also unknown unknowns. There are things we do not know we don't know." We may set out to produce information about the known unknowns, but the unknown unknowns are likely to cause us unexpected trouble – how are we to investigate when we do not know what we do not know? And on the other side, there are the words attributed to humorist Josh Billings: "The problem with people ain't ignorance – its things folks know that ain't so."(1) If we think we know the answer, then we are not likely to invest in learning the truth. And some people who have information we need may have reasons to conceal it or misrepresent it. Thus, all in all, the production of information can be an unpredictable process.

Thus, from this point on in the book we will take it for granted that information is never free. In some small groups, this may be a rather minor problem. In the McCain and Sons farming partnership, which in the early 1950s comprised my father, grandfather, and uncle, it was the custom of the partners to meet in the evenings at my grandfather's home to plan and coordinate their activities for the next few days. My grandfather would consult his journals to recall what had been done in previous

* Information may also be a first order good, whenever people have wants or needs for information, but here we are concerned with information that complements other higher order goods in increasing the value of the coalition.

years, and how successful it had been, and the tasks to be done would be apportioned. This commitment of time was not negligible, but my grandfather and his sons enjoyed their time together (usually!), so that it was not altogether a cost.

In many other situations, however, the information necessary to achieve the very best allocation of tasks and resources, and even to find the information we might use to discover the best allocation, can be quite costly. In these circumstances, there are a number of arrangements that may reduce the need for detailed information or the costs of obtaining the information, at some sacrifice in the gross value created by the coalition. The coalition may gain more by reducing the cost of information than it loses through a less-than-perfect coordination of tasks and resources, so that the value created is increased on net by being less informed. Among the arrangements that conserve on costly information are hierarchies of authority and standing rules of procedure. Together, hierarchies of authority and established rules of procedure are called governance. Relying on governance, we may not realize the very greatest value in every instance. Nevertheless, after allowing for the costs of information, the overall average value creation of the group may be greater when it relies on governance than it would be if the group obtained the information necessary to bring about the very best course of action in every case. A coalition that relies on governance to determine much of its day-to-day activity may be called an organization.

In this chapter we will illustrate, by example, some of the many ways that the cost of information influences economic organization and explains some of the most familiar aspects of the economic environment. In the next chapter we will return to game theory to consider how and why hierarchies and rules of procedure can economize on information costs.

THE COST OF TRANSACTION[(2)]

As we observed in the first chapter, exchange is one of the ways that people join together for mutual benefit. Exchange is not a particularly simple activity! Let us think of exchange as a game of strategy. When I buy a cup of latte from the Starbucks at Drexel University, my "strategy" is to give them a certain amount of money. Starbucks's "strategy" is to prepare and give to me a cup of caffè latte. Neither of us would benefit by adopting our respective strategies unilaterally: it is the reciprocity, the transaction by which we adopt complementary strategies, that makes the exchange mutually beneficial. Any explicit agreement to adopt "strategies" or courses of action that lead to mutual benefit is a transaction, for the purposes of this book.

There is a lot of information that is reflected in my morning transaction with Starbucks. Starbucks knows how to make a caffè latte of a certain quality and kind, and on the basis of experience I know that I like a caffè latte of that quality and kind. I know where the Starbucks counter is located and when it is open. I know the price I will be asked to pay. There is one "known unknown": until I get there I do not know how long the queue will be. Sometimes it is just too long, and so I do not get a latte on that particular day. Standing in the queue is an example of a cost of transaction. If, on a particular day, my willingness to pay for a Starbucks caffè latte is greater than the cost of producing it, then there is a potential mutual benefit, a surplus of value over cost, for me and Starbucks if we carry out the transaction. But standing in the queue is a use of my time that I would rather avoid, and thus a cost to me. If the cost of transaction, standing in the queue, is greater than the surplus, then the transaction will be a loss overall – inefficient – so it is better that the transaction does not take place.

A contract of employment between an employer and an employee is another instance of a transaction. The employee accepts the contract because it is better than his or her other opportunities, and the employer accepts it because the candidate has the skills he or she needs for a profitable operation. Both are likely to spend some resources finding out about the alternative opportunities before they commit themselves to the contract. The employer may review and interview a number of candidates, a time-consuming and costly process. The employee will also have spent time going to interviews and filling out applications. The employee may not accept the first offer but remain in the search until he or she is satisfied that the offer he or she has is the right one to accept. If the employee has declined an early job offer he or she has given up the option of working at that job, even if it is on worse terms, for the time that elapses between the two offers. This, too, is a cost. The employee and the employer are willing to undergo these costs because they expect that their transaction will produce a surplus more than enough to offset the costs of transaction.

In the employment example, the transaction costs are the costs of information. Suppose that information were free. Then both parties would gather all the information that might be helpful to them. The employee would know exactly the best offer they could hope for, and would accept nothing less; the employer would know exactly who is available, their skills and what they would accept, and would immediately make the best offer to the best employee. However, because information is costly, there is a limit to the information they will gather. If they gathered all the information that would be helpful to them, the cost would be far more than the surplus arising from their transaction – both would be worse off, not better. So they stop gathering information at some point, make a decision based on what they do know, an offer and (the employer hopes) an acceptance.

This suggests that costs of transaction are very important with respect to employment and labor, and research honored by the 2010 Nobel Memorial Prize in Economics demonstrates that indeed they are.(3) In the employment example, the costs of transaction arise from the cost of information. Can we apply that to the Starbucks example? It may not appear that the cost of standing in line at Starbucks is a result of costly information, but when we think it through carefully, it really is. Consider: many restaurants accept reservations, so that no one needs to stand in line to wait for a table. Starbucks does not. In principle, they might. My observation suggests to me that the Starbucks location I usually buy from can process a transaction in about 30 seconds. Suppose that each customer had an assigned time slot of 30 seconds at about the time she or he wants coffee, and each person showed up just on time. No one would have to stand in line – but the informational cost of doing this would be stupendous. The informational cost of 30 second reservations at Starbucks would be far greater than the cost of people standing in line, and that is why people stand in line – because information isn't free! By contrast, when I make a reservation for dinner at a fine restaurant, I expect to remain at the table for an hour or more, spend a great deal more money, and am likely to plan ahead and so it is natural enough to call ahead for a reservation. These restaurants have fewer patrons, stand to lose more if tables are vacant, and so find it less costly to accept the call and record the reservations (that is, produce information), rather than simply serving those who walk in as Starbucks, and some other restaurants, do. For a restaurant of this kind, and for its customers, the information cost of individual reservations is real and important, but less than the cost of having customers queue for tables. And this illustrates an important point. Businesses and their customers and employees will adapt their decisions to the costs of information, trading off informational costs against other costs, but they will trade those costs off in different ways depending on their different conditions.

INCOMPLETE CONTRACTS

A transaction can be very simple. At a bazaar or farmers' market, the money in my hand is exchanged for some of the fresh beans at the farmer's booth, and that is an end to it. Other transactions can be quite complex, and can continue over a considerable period of time. Consider, for example, an agreement between a landowner and a professional forester to manage a woodlot owned by the landowner. The forester will harvest timber for lumber or paper pulp, sell it, retain a fee, and pay the balance to the landowner; will replant and take other needed action to prepare the woodlot for another harvest in a few years. The agreement between the two will need to specify the fee, and which trees are to be harvested: is the woodlot to be clearcut or will the larger trees be harvested, and the rest only thinned for sustainable harvesting? If sustainable harvesting is chosen, how large must the trees be to be harvested: is a 10 inch circumference big enough, or should it be 12 inches? In addition the agreement must specify whether costs of replanting and administration will be paid from the forester's original fee or whether the forester will bill the landowner separately for them. Some standards must be set for the administration. The term of the agreement may be measured in decades. In these circumstances, a written contract will be essential in order to assure both parties of the benefits they can expect from the agreement.

A contract is, from one point of view, a body of information. The information in the contract is a description of what the parties to the agreement have agreed to. These agreements may pertain to things that might never happen. Again, consider the contract between the landowner and the forester. Suppose, for example, that oil or natural gas should be discovered in the region. Drilling for oil or natural gas could interfere with the management of the replanted forest. In northwest Louisiana – where my woodlot actually is – drilling for oil and gas is fairly common and it would be smart for a forest management contract in this area to include some agreement on how this interference would be handled. Drilling might be prohibited, or, more probably, an adjustment in the terms of the management contract might be specified. Despite the agreement, it very well might happen that there is no drilling. An event that may or may not happen during the term of the contract, such as the discovery of oil or natural gas in the area, is sometimes called a contingency. In complicated transactions many contingencies will be dealt with in the contract.

If information were free, the contract could cover every single contingency that the parties to the contract can envision. However, to repeat the theme of this chapter, information is not free. Because it is not, the cost of a contract that covers every contingency is likely to be more than the surplus value created by the agreement. As a result, some contingencies will not be dealt with in the contract. In an area in which drilling for oil or natural gas is unusual, and there is little probability that such resources will be discovered, the parties to a forest management contract might not deal with that contingency at all. It would not be worthwhile to spend the time in discussing and settling the adjustments to be made for a contingency that is very improbable. Thus, contracts for many complex transactions will be incomplete. Some foreseeable contingencies will be left out, because they are not likely enough to justify the use of resources to extend the contract to deal with them. In addition, we have what Secretary Rumsfeld called "unknown unknowns." Whether oil or natural gas may be discovered is a known unknown. The contract may or may not deal with it, depending on the costs of doing so and the risks of not doing so. However, no contract can deal with contingencies that the parties to the agreement do not envision, the unknown unknowns. Perhaps, if information were free, there would be no unknown unknowns – perhaps we could then produce enough information at least to know what all the unknowns are – or perhaps not. In any case, unknown unknowns play a part in many agreements, and the contracts for those agreements are unavoidably incomplete.

Economists have only studied incomplete contracts and their implications since the 1960s.(4) (There may be some exceptions but they certainly are not well known.) There is still a great deal to be learned. Nobel Laureate Oliver Williamson has done particularly important work on the implications of incomplete contracts. It is clear, as he has stressed, that the incompleteness of contracts can explain some predicable features of the contracts people do agree to. Sometimes, the parties to a contract will face circumstances not dealt with in their incomplete contract, such as oil or gas discovery and drilling in an area where it has not been expected. It may then be necessary to renegotiate some parts of the contract. This renegotiation will be costly and there is the risk of disagreement disrupting the previous agreement, even of litigation. These risks have to be traded off against the cost of making the contract more complete. In some cases, the contract may anticipate the unanticipated by establishing rules to deal with the unexpected, or to limit and guide renegotiation. A contract clause requiring arbitration would be an example of this kind. We will find that there are many others.

Incomplete contracts are particularly important in agreements between an employer and an employee. These transactions are often long-lasting, even to the point (in a few cases) of lifetime jobs. For this and other reasons, it would be particularly difficult and costly to make these contracts complete. The parties to the contract are not likely to attempt to anticipate every task that the employer may need to have the employee complete. Thus, many employment contracts contain a clause that says the employee is responsible for "other duties as assigned." If the employer is a business proprietor, any unexpected gains to the surplus created by the agreement will belong to the employer as proprietor. In larger firms, these uncontingent arrangements will be extended and specified to such a degree that we recognize the firm as a form of organization.

IMPLICIT CONTRACTS

In the previous section we discussed cooperative agreements based on written contracts and noted that, since information is not free, such contracts will often be incomplete. In some cases there may be no written contract at all. One thing that we learn from non-cooperative game theory is that, if a relationship is long-lasting, the parties to the relationship may be able to achieve cooperative strategies without an explicit (spoken or written) agreement.(5) This idea is most often applied to cartels. A group of firms compete on price and can increase their profits if they can agree cooperatively to charge a (high) monopoly price. The strategies for their "game" are to charge a monopoly price or to cut price. If a firm cuts price when the others do not, then its profits will increase above the monopoly level, as it takes customers away from the other firms in the group; but if all follow this reasoning and cut the price, then their profits will decline below the monopoly level. Charging a monopoly price is a cooperative agreement among the firms, which enables them to extract more profit at the expense of the customers. In this it is similar to the bidding ring in the example of the auction. The firms will continue to play this pricing game over an indefinite future. However, an agreement to charge the monopoly price would be illegal. Nevertheless, each firm might restrain itself from cutting the price because of an expectation that another firm would retaliate by cutting the price in turn on a future round of play. Thus, the monopoly price might be maintained even though there is no agreement to do so.

Something of the kind could also occur in relations between an employee and an employer. The employee would prefer to make as little effort as possible, no more effort than is necessary to keep the job. The employer would prefer to cut the wage to the lowest wage that would retain the worker. However, both can be better off if the employee makes an efficient effort and the employer pays a somewhat higher wage. If the employer–employee relationship is persistent, then the employee may continue to make an efficient effort, expecting that the wage advantage will soon disappear if he or she does not, and the employer may continue to keep the wage above the employee's best alternative, understanding that the employee will withdraw the extra effort if he or she does not. All of this takes place without a written contract or oral agreement. In twentieth-century economics unspoken contracts between employers and employees have been called implicit contracts.* The influential economist Arthur Okun stressed their importance and, with a bit of Smithian wordplay, referred to the implicit contract as "the invisible handshake."(6)

Implicit contracts may be important in other ongoing relations as well. It would seem that implicit contracts have some disadvantages by comparison with explicit, written contracts.** Since no effort has been made to identify contingencies and to agree on the common response to them, even for contingencies that are moderately probable, the agreement may break down if the contingencies occur. Since there has been no negotiation in the first place, renegotiation in unusual circumstances seems unlikely. All in all, an implicit contract can deal only with normal conditions, and is likely to break down in the sort of abnormal circumstances that we will sometimes experience.

Why, then, would people settle for an implicit contract rather than an explicit and written one? Once again, the answer is: information is not free. While normal times continue, an implicit contract requires very little information at all. Thus, because the informational costs of negotiating, signing, and monitoring a written contract are avoided, the net surplus available to the parties to the relationship are increased, and both are better off – at least until the normal times give way to abnormal circumstances.

Written and implicit contracts may coexist in a particular relationship, and here, again, relationships between employers and employees offer examples. An example will be given in Chapter 9, so for now we will only remark that the employer and employee may well achieve a cooperative arrangement, with efficient effort and wages and working conditions better than the market minimum, even in the absence of any written or spoken agreement to act cooperatively on these matters, another instance of the "invisible handshake."

The key points are that, when the contract is incomplete or there is no contract at all, non-cooperative action may fail to realize opportunities for mutual benefit. If the relation is long-lasting, however, the cooperative strategies may be realized on the basis of an unspoken implicit contract. Such an "invisible handshake" is important in many employer–employee relationships and may be important in many other ongoing relationships among people who work together for mutual benefit.

* If the relationship is between an employer and a group of employees, then things are a bit more complicated, in that the relations among the employees may also be important. Nevertheless, the economic theory of implicit contracts is also applied to the employment of large groups, and, in practice, this complication does not seem to be a problem.

** What follows in this paragraph is conjectural, as the topic does not seem to have been studied in economics.

ORGANIZATION

In this chapter so far, we have put a good deal of emphasis on groups and agreements that persist over time. Some of these coalitions can be described as organizations. An organization is a coalition that (1) continues over time, usually for an indefinite term, (2) includes more than two people, and usually a large number, (3) within which many decisions are predetermined or influenced by rules, which are also persistent, and (4) within which there is a hierarchy in which those at higher rank will sometimes direct the decisions or strategies or those who are below them in the hierarchy. Among common examples of organizations are business corporations and other medium-to-large-scale business enterprises, government agencies, military units, and many non-profit, religious, and other groups.

The rules, procedures, and hierarchy within an organization constitute its governance, and the next chapter will take up governance in detail. For now, we merely remark that the persistent rules and hierarchy are means of economizing on costly information. If information were free, all of those active in an organization might well sit down together and work out their common plan, as my father, my grandfather, and my uncle did for their business, but, for reasons the next chapter will explore, many businesses and other groups engaged in common action will find it more convenient to rely on organization.

EXTERNALITY

In economics, an externality is a case in which a (non-cooperative) decision taken by one agent imposes costs or benefits on other agents. As an example, consider a well drilled to obtain water from an underground aquifer, that is, an underground pool or stream that extends over a considerable area. Whatever water is withdrawn by one person living in the area lowers the level of the aquifer and thus makes it more difficult and costly for others to obtain water from the aquifer. Since the individual well-driller does not take into account the costs to others (when the decisions are made non-cooperatively), too many wells are drilled, the aquifer is drawn down more than would be most productive, and the total surplus of benefits over costs is less than it might be, and might even be negative. This is an instance of a negative externality.

We may also see examples of positive externality. In some circumstances, education may give examples. For example, where some farmers are better educated, their education enables them to make better decisions, but these decisions are imitated by other farmers in the district, so that even the less-educated farmers are more productive as a result of the higher average level of education in their district.[7] But each individual, making decisions non-cooperatively, considers only the benefits of his or her own education, and so not enough investment in education is undertaken.

As we see, in a non-cooperative framework, externalities can lead to inefficiencies. Why, then, do the property owners in the area of the aquifer, or the young people aspiring to farm in the district, not form a cooperative coalition and commit themselves to efficient use of the water from the aquifer and to an efficient rate of investment in education? The reason is transaction costs. In each case, an efficient response to the externality would require a coalition among a large number of people, and the costs of information to determine what commitments each person should make, to monitor and enforce the agreement, and generally to carry out the transaction, could be quite high. Thus, if the costs of transaction are greater than the surplus realized by the coalition, the coalition will not take place, and the inefficiencies will be observed.[2]

One important qualification needs to be made. We have seen that rules of procedure can economize on the costs of information and they may make a cooperative solution feasible where it otherwise would not be. With respect to externalities, the rules and procedures may be of two kinds. One kind is government regulation. This will be discussed in Chapter 8. The other kind is reliance on customary norms. Customary norms can be thought of as instances of an implicit contract, since they are not (originally) written down but nevertheless function somewhat as a contract among the people affected by the externality. Customary norms can be effective in increasing the efficiency of management of externalities, and in particular the efficient use of water supplies seems sometimes to have been accomplished in this way.[8]

MONEY

An old tradition in economics explains the function of money as a means of conserving costly information. This takes the form of a contrast between money exchange and barter. The argument is that, in the absence of a monetary system, people would have to rely on barter to enter into mutually beneficial coalitions for exchange. But barter cannot take place between two people unless there is double coincidence of needs.[9] That is, for example, if an auto mechanic needs a haircut he must find a barber who needs his car repaired. (This is a relatively easy example! If *I* need a haircut, I would have to find a barber who wants a lecture on economics – and I suspect that there are many more barbers who need a car repaired than who want an economics lecture.) At the very least, there will be an informational cost of finding that barber. In a money economy, the auto mechanic with money in his pocket needs only find a barber. There will be an informational cost of doing that, but the informational cost of finding a barber *specifically* who needs his car repaired will be much greater.

Of course, bilateral exchange is not the only possibility. It might be that the auto mechanic with the shaggy hair could find a barber who needs some dinner rolls and a baker who needs his car repaired, and arrange for a three-way exchange; or find a barber who needs dinner rolls, a baker who needs a new jacket, and a tailor who needs his car repaired, and arrange for a four-way exchange. But the informational costs must be going up at least in proportion to the number of people involved in the exchange. In our imagination, we might extend this process until we have a multi-way exchange among the whole population – a perfect central plan for a war communist* economy. But it is difficult to imagine the informational cost of actually doing that.

In practice, in a barter economy, very little exchange will take place, since the informational cost of an exchange will very often be greater than the surplus created by the exchange. Jevons gives an example[(9)]

> When Mr. Wallace was travelling in the Malay Archipelago, . . . he tells us that in some of the islands, where there was no proper currency, he could not procure supplies for dinner without a special bargain and much chaffering upon each occasion. If the vendor of fish or other coveted eatables did not meet with the sort of exchange desired, he would pass on, and Mr. Wallace and his party had to go without their dinner.[(10)]

In practice, our auto mechanic will probably cut his own hair.

BOUNDED RATIONALITY

Much economic theory begins from the idea that consumers spend their money so as to maximize utility, and that business people make their business decisions so as to maximize profits. In the words of game theorist Reinhard Selten, this sort of economics and game theory is about "absolutely rational decision makers whose capabilities of reasoning

* In the early days of the Soviet Union, during the Russian Civil War, the Soviet Union experimented briefly with an economic system without any prices at all, by directly allocating consumer goods and tasks to individuals throughout the part of the country under their control. This was called "war communism" and is the origin of the phrase. It seems fair to say that war communism failed very rapidly, and after its abandonment it was said to have been a wartime emergency measure.

and memorizing are unlimited."[11] If the real world were inhabited by such creatures, then information would be free, since each person could produce an infinite amount of it at no cost. (There might still be some difficulty with the unknown unknowns, if any were left after an infinite quantity of information had been produced.) To say that we live in a world of costly information, then, is to say that we live in a world in which decision-makers are not absolutely rational but boundedly rational. That is, there is a limit to the information that a real human being can produce, at any limited cost. Real rationality is bounded rationality. That phrase and much of our understanding of bounded rationality come from Herbert Simon,[12] but the point has also been stressed by Oliver Williamson and Harvey Leibenstein, among many others.

That does not in itself mean that people do not maximize. However, in order to maximize utility or profits, it will be necessary to discover what quantities of consumer goods or inputs and outputs correspond to the maximum, and in practice also how they must be used, with what timing and strategies, who may be most suitable as partners, employees, customers, and suppliers, and many other details. That is a lot of costly information. In a few simple cases, people may maximize, but in general they have good reason to look for some way to conserve this costly information.

As we have seen, rules may conserve information. If we can compile a set of rules of procedure that can be applied with only a little, cheap information, and that will usually *approximate* the maximum, then we may be better off with the simple rules than we would be if we produced more information to get a better approximation. A rule of procedure that usually approximates the maximum utility or profit is called a heuristic rule, or sometimes a rule of thumb. These heuristic rules play a key part in Simon's conception of bounded rationality. This is especially clear in Simon's work after 1970, in which, working with Alan Newell, Simon was a founder of the computational field called artificial intelligence.[13]

In an essay entitled "Rules of thumb and optimally imperfect decisions," Baumol and Quandt[14] described the kinds of rules that make sense as guidelines for good business decisions:

1. The variables that are employed in the decision criteria are objectively measurable.
2. The decision criteria are objectively communicable, and decisions do not depend on the judgment of individual decision-makers.
3. As a corollary to number 2, every logically possible configuration of variables corresponds to a (usually unique) determinate decision.
4. The calculation of the appropriate decision is simple, inexpensive, and

well suited for frequent repetition and for spot checking by management in higher echelons.

However, there are some problems with the idea of optimally imperfect decisions, as Baumol and Quandt acknowledge. An optimally imperfect decision is one that maximizes the *net* utility or profits of the decision-maker, after allowing for the cost of producing information. Simon dismissed the idea as meaningless, saying that if people are not capable of discovering the strategy that maximizes their payoffs, then they will be even less capable of solving the more complex problem of determining just how much information to produce in order to discover it. Simon argued instead that as long as the outcome of the existing rules of thumb are satisfactory, there will be no attempt to improve them. This has been called a theory of satisficing as opposed to maximizing decision-making. In this, Simon drew on psychological thinking of the time (and since) that argued that people often have fairly well-defined aspirations, and that if their decision rules satisfy those aspirations, then there will be no decisive motive to change them.

Most economists dislike Simon's "satisficing" theory because it simply does not answer the questions the economists want to ask. Some economists have observed that aspirations are not just constant, and are sensitive to experience, so that we might envision an evolution-like learning process in which aspirations rise when there are plenty of opportunities, so that the satisficing decisions gradually move in the direction of optimally imperfect decisions. This sort of thinking has been called adaptive economics.(15) However, there is much to learn, as only a little research was devoted to adaptive economics. One thing we do know is that an economy in which decisions are optimally imperfect will not behave at all like one in which information is free and decisions are perfectly rational.(16)

A different position that was much more popular among economists went in just the opposite direction and dominated economic research in the last quarter of the twentieth century. It was called rational expectations. The contrast deserves a bit more comment. In order to make decisions either as a business person or in one's personal life, one needs two kinds of information. One kind of information describes the impact of one's own strategies on one's payoffs, and the other kind of information describes the impact of other events, including both natural events and the decisions of other persons, and what those events are likely to be. The latter kind of information is known in economics as expectations. Outside economics, many people may be skeptical about the economist's idea of rationality as maximization, but would nevertheless agree that a rational person would base his or her expectations on the best available evidence and reason, and that this could differ from simple adaptive learning. The theory of rational expectations put this idea to work within economics.[17]

The theory of rational expectations was more truly a response to the ideas of John Maynard Keynes than to those of Simon. It was Keynes who made economists aware of the importance of expectations,[18] since he held that "the state of business expectations" is an independent influence on investment decisions and many other business decisions. Some rational expectations theorists have suggested that Keynes just neglected the rational basis of expectations, but that would be a distortion of Keynes's ideas. It is pretty clear that Keynes didn't think that anything like rational expectations could be possible – he didn't think that people could ever obtain enough information to form rational expectations. In the Rumsfeldian terms we have been using in this chapter, perhaps Keynes would agree that there are simply too many unknown unknowns – and he would say that business people know that there are.

The truth is probably somewhere between Keynes and the rational expectations theorists. That middle ground is the territory of adaptive economics, and one may hope that future research in the economics of expectations and decision-making could be focused on that middle ground.

ENTREPRENEURSHIP

In popular discussions of economics, the issues often seem to turn on the role of the entrepreneur. In Chapter 3, using the language of American economist John Bates Clark, we described the coordinator for a coalition as an entrepreneur. In the research literature of economics, the entrepreneur also gets a lot of attention, but a difficulty is that there are many different definitions and concepts of the entrepreneur and entrepreneurship in our research literature. For the purposes of this book, it will be best once again to return to the elder Austrian School of thought and begin with the ideas of Karl Menger. Menger writes:

> Entrepreneurial activity includes: (a) obtaining information about the economic situation; (b) economic calculation—all the various computations that must be made if a production process is to be efficient (provided that it is economic in other respects); (c) the act of will by which goods of higher order . . . are assigned to a particular production process; and finally (d) supervision of the execution of the production plan so that it may be carried through as economically as possible.[19]

We can recognize a, b, and d as information production in the way we have used the term in this book. Menger also identifies entrepreneurship as a labor service and a higher order good, but notes that unlike many other higher order goods, entrepreneurship is not a service that is bought and sold in markets. (We have seen that there can be some obstacles to markets for buying and selling information.) Menger also recognizes that some of these functions can be delegated to assistants. In this context, item c – the act of will to set the process of production in motion – stands out as an exception that could hardly be delegated.

Probably the best-known economist in the discussion of entrepreneurship is Joseph Alois Schumpeter.(20) His understanding of the word is quite different from Menger's. Schumpeter regards as entrepreneurs only those who introduce innovations, and within that group, only those whose innovations are history-making – the Henry Fords and Steve Jobs of economic history. (Schumpeter stresses, though, that "innovation" means more than technological innovation, and may be organizational or market innovation, including the creation of a new monopoly.) In short, Schumpeter identifies the entrepreneur with "the creative response," and more specifically with those instances in which the creativity is great enough to generate a break from the routine of the market economic process.

To keep the senses of the word "entrepreneur" distinct, from this point in the book the term will be "creative entrepreneur" when the Austrian or Schumpeterian concepts are meant, and simply "entrepreneur" when John Bates Clark's sense is meant. A Clarkian entrepreneur may or may not also be a creative entrepreneur, but (as we will see in the next chapter) a creative entrepreneur will always be a Clarkian entrepreneur. The difference is found (in Schumpeter's words) in "the creative response." The Clarkian entrepreneur may always respond to opportunities and challenges adaptively, but if he or she responds creatively, then he or she is a creative entrepreneur – this seems a natural enough distinction.

Even so, we have one phrase, "creative entrepreneur," used to denote two concepts, and both are important concepts. Schumpeter's ideas are almost unique in explicitly treating creativity as an aspect of the economic process, and that is a very important insight. But in limiting the role of "entrepreneur" to those whose economic creativity is history-making, it may be that Schumpeter pushes the creative use of language a bit too far. Most business founders do not expect to create the next $50 billion corporation.(21) However, the two concepts are linked by Menger's item c – the initiative, the "act of will," the decision to form a new coalition, or to renew and transform an existing one. This certainly is the beginning point for Schumpeter's historic innovators. But one should not underestimate the creativity necessary to found a local hairdresser's salon, butcher shop or dairy farm and to keep it operating in a world in which unknown unknowns occasionally make themselves known. And modern thought on entrepreneurship stresses that the creativity that renews a business does not always come from the top but that "intrapreneurship" and "social entrepreneurship" are increasingly important in real, advanced economies.[22] On the other hand, it seems that to be a historic innovator, an entrepreneur in Schumpeter's sense, it will often be necessary first to be an entrepreneur in Menger's sense, and in that of John Bates Clark.

In a world of costly information, the entrepreneur's task is more than just the recognition of new opportunities and acting on them. The entrepreneur who initiates a new coalition will have to create the new routines and rules of thumb, forms of delegation, and organization necessary to realize the opportunity for mutual benefit among those who are joined together in the organization. In a world of free information all this might be trivial, but information is not free.

CHAPTER SUMMARY

Information is a higher order good needed along with other higher order goods to produce goods and services to meet individual wants and needs, and to engage in transactions between the producers and the consumers and to form other coalitions for mutual benefit. Like other higher order goods, information is costly to produce. This fact explains some of the most prominent facts of economic life, including money, bounded rationality, incomplete and implicit contracts, unemployment, inefficiencies created by externalities, and organizations with their hierarchies and rules of procedure. The world we experience is imperfect in many ways. Because people are both creative and adaptive and can learn and create new routines, we may see a movement toward an optimally imperfect organization of the economy. We might hope that this optimally imperfect organization would resemble the equilibrium in a world of perfectly rational beings, but this does not seem to be the case. To understand the imperfect economic world we live in, we will have to understand the imperfections.

SOURCES AND READING

The role of information and information costs in economics has been the subject of not one, but a number of large, disjoint literatures in economics since the mid-twentieth century and this chapter will draw selectively on a number of them. (1) Rumsfeld's comments were made on 12 February 2002 at a press briefing, according to Wikipedia, http://en.wikipedia.org/wiki/There_are_known_knowns, as of 3 August 2012. On that date a

YouTube video of the comments was available at http://www.youtube.com/watch?v=_RpSv3HjpEw. The comment has widely been regarded as a "political gaffe" but a few have seen it as insightful, and it is the insight that is valuable for our purpose. Billings seems actually to have written: "It is better to know nothing than to know what ain't so" (Wikiquotes, http://en.wikiquote.org/wiki/Josh_Billings, accessed 6 February 2011). Elaborations like the one in the main text have often appeared in print and online, with or without attribution to Billings, but in any case he does seem entitled to credit for the idea.

(2) Almost all discussions of transaction costs can be traced back to Ronald Coase (1937), "The nature of the firm," *Economica*, New Series, **4**(16), 386–405. Ronald Coase (1960), op. cit., Chapter 1 of this volume, note 10, was even more influential in its time, though perhaps less so today. In any case, no one has done more to make economists aware of the importance of costs of transaction than Ronald Coase, who was honored by the Nobel Memorial Prize for his ideas in 1991. However, Coase was not aware of game theory, and, conversely, most of the literature of cooperative game theory has assumed away informational costs. Nevertheless, one key idea of Coase's (1960) was shown to be incomplete from the viewpoint of cooperative game theory: see Varouj A. Aivazian and Jeffrey L. Callen (1981), "The Coase theorem and the empty core," *Journal of Law and Economics*, **24**(1) (Apr.), 175–81. All the same, without Coase's influential work, our understanding of the importance of transaction costs would not exist. The discussion of externality in the fifth section below directly follows Coase (1960) in connecting externalities to transaction costs. The work of Oliver Williamson is no less important for our understanding of transaction costs. It was honored by the Nobel Memorial Prize of 2009. See especially Oliver E. Williamson (1975), *Markets and Hierarchies*, New York: Free Press. This work unites a focus on transaction costs, incomplete contracts, bounded rationality, and opportunism, and addresses forms of organizational governance as an explanandum. Thus, it has many parallels with the ideas put forward in this chapter and the next, and accordingly some further reference will be made to it there. Further reference to Coase and Williamson will be made in respect of some particular points in their work.

(3) See, for example, Dale T. Mortensen and Christopher Pissarides (1994), "Job creation and job destruction in the theory of unemployment," *Review of Economic Studies*, **61**(3), 397–415, and many other publications referenced on the Nobel Prize site. We will return to this in Chapter 9. (4) The early work of Oliver Williamson (e.g., op. cit., 1975) on incomplete contracts is very important and is the main source of further work. Harvey Leibenstein's earlier work also stressed the importance of incomplete

contracts. See Harvey Leibenstein (1969), "Organizational or frictional equilibria, x-efficiency, and the rate of innovation," *Quarterly Journal of Economics*, **83**(4) (Nov.), 600–23. Leibenstein particularly influenced the author of this book. (5) Technically, the cooperative strategies discussed in the context of repeated play correspond to a "subgame perfect Nash equilibrium" in a game played repeatedly without a definite time limit to the play. A detailed discussion is beyond the scope of the book; see Roger A. McCain (2014, Chapter 15), op. cit., Chapter 1, note 8 of this volume, for example, for a more extended discussion. The condition of subgame perfection somewhat limits the cases in which cooperative strategies will be played, but experiments suggest that real human beings can sometimes cooperate without explicit agreement even when these restrictive conditions are not met. (6) Okun (1928–80) used the term "invisible handshake in the title of a paper published in *Challenge* magazine: Arthur M. Okun (1980), "The invisible handshake and the inflationary process," *Challenge*, **22**(6) (Jan./Feb.), 5–12. The ideas were worked out in more detail in his 1981 book, *Prices and Quantities: A Macroeconomic Analysis*, Washington, DC: The Brookings Institution, published posthumously. The literature on implicit contracts began with Costas Azariadis (1975), "Implicit contracts and unemployment equilibria," *Journal of Political Economy*, **83**(6) (Dec.), 1183–202.

(7) While the educational externality has not been much investigated, it was a finding in George Fane (1975), "Education and the managerial efficiency of farmers," *Review of Economics and Statistics*, **57**(4) (Nov.), 452–61. Most studies of the impact of education on productivity have looked at *either* individual *or* regional (i.e., county) averages, and these studies would not give information on the positive externality suggested here. (8) Elinor Ostrom (1933–2012) who shared the 2009 Nobel Memorial Prize with Oliver Williamson, is the source of much of our knowledge of the role of customary norms in managing externalities. See the Nobel website, http://www.nobelprize.org/nobel_prizes/economic-sciences/laureates/2009/ostrom-facts.html, available as of 30 July 2013, for further information. (9) The phrase "double coincidence of needs," and the explanation of the function of money in terms of it, seem to originate with W. Stanley Jevons, 1876, although the idea was clearly expressed in Smith's *Wealth of Nations* and was probably old then. See W. Stanley Jevons (1876), *Money and the Mechanism of Exchange*, New York: D. Appleton and Co. (first published 1875) available at http://www.econlib.org/library/YPDBooks/Jevons/jvnMME.html, as of 15 August 2012. (10) Jevons (op. cit., Chapter 1, para. 2). While Jevons is not explicit, presumably he refers to Alfred Russel Wallace, the co-discoverer with Darwin of the theory of evolution, whose expeditions provided confirming evidence for the theory.

(11) Selten's quotation is from Reinhard Selten (1975), op. cit., Chapter 1 of this volume, note 9. (12) Herbert Simon has discussed bounded rationality on several occasions. See especially Herbert A. Simon (1955), "A behavioral model of rational choice," *Quarterly Journal of Economics*, **69**(1) (Feb.), 99–118, and Herbert A. Simon (1959), "Theories of decision-making in economics," *American Economic Review*, **49**(3), 253–83. Simon's ideas were honored by the Nobel Memorial Prize in 1978 and his Nobel lecture, "Rational decision making in business organizations," is probably the best reading that puts bounded rationality in the context of his complex, interdisciplinary system of ideas. It is available at http://www.nobelprize.org/nobel_prizes/economics/laureates/1978/simon-lecture.pdf, as of 21 August 2012. (13) For an overview of Simon's ideas on artificial intelligence, see Herbert A. Simon (1995), "Artificial intelligence: An empirical science," *Artificial Intelligence*, **77**(1), 95–127. (14) The "optimally imperfect decisions" criteria are from William J. Baumol and Richard E. Quandt (1964), "Rules of thumb and optimally imperfect decisions," *American Economic Review*, **54**(2) (Mar.), 23–46 at p. 24. Although they do not cite it, one of the most important instances of optimally imperfect decisions was given by George Stigler (1961), "The economics of information," *Journal of Political Economy*, **69**(3) (June), 213–25, a paper that founded the literature on search and matching in economics and that was honored by the Nobel Memorial Prize in 1982. Another important case in the recent literature is Christopher A. Sims (2003), "Implications of rational inattention," *Journal of Monetary Economics*, **60**(3), 665–90, which proposes that people put optimal limits on the sources of information that they monitor. (15) On adaptive economics, see, for example, Richard H. Day (1967), "Profits, learning, and the convergence of satisficing to maximizing," *Quarterly Journal of Economics*, **81**(2) (May), 302–11; John G. Cross (1973), "A stochastic learning model of economic behavior," *The Quarterly Journal of Economics*, **87**(2) (May), 239–66; Richard H. Day (1978), "Adaptive economics and natural resources policy," *American Journal of Agricultural Economics*, **60**(2) (May), 276–83; John G. Cross (1983), *A Theory of Adaptive Economic Behavior*, Cambridge: Cambridge University Press; and for a more recent, explicitly evolutionary approach, Giovanni Dosi, Luigi Marengo, Andrea Bassanini and Marco Valente (1999), "Norms as emergent properties of adaptive learning: The case of economic routines," *Journal of Evolutionary Economics*, **9**(1) (Feb.), 5–26. Adaptive mechanisms have also played a role in many studies using the method of agent-based computer simulation. On this method see Leigh Tesfatsion (2006), "Agent-based computational economics: Growing economies from the bottom up," Department of Economics, Iowa State University, available at http://www.econ.iastate.edu/tesfatsi/

ace.htm, as of 21 August 2012, last modified on 2 May 2012. (16) The discussion of the failure of boundedly rational equilibria to approximate rational ones follows George A. Akerlof and Janet Yellen (1985), "Can small deviations from rationality make significant differences to economic equilibria?" *American Economic Review*, **75**(4), 708–20.

(17) Rational expectations economics arose from a paper by John F. Muth (1961), "Rational expectations and the theory of price movements," *Econometrica*, **29**(3) (July), 315–35. It seems no coincidence that the key papers of Stigler and Baumol and Quandt, mentioned above, and of Muth, all followed soon after Simon's 1959 paper, though it might be an exaggeration to describe them as responses to it. Several rational expectations theorists have been honored by the Nobel Memorial Prize, notably Sargent (2011), http://www.nobelprize.org/nobel_prizes/economics/laureates/2011/sargent.html; Phelps (2006), http://www.nobelprize.org/nobel_prizes/economics/laureates/2006/; and Lucas (1995), http://www.nobelprize.org/nobel_prizes/economics/laureates/1995/. All web addresses were active on 21 August 2012.

(18) The determining role of expectations was put forward in John Maynard Keynes (1997), *The General Theory of Employment, Interest, and Money*, Amherst, New York: Prometheus Books, originally published in 1936. However, the implications of Keynes's position cannot really be understood without reading his earlier work on uncertainty, John Maynard Keynes ([1921] 1973), op. cit., Chapter 5 of this book, note 1. Keynes's ideas on uncertainty were closer to those of his American contemporary Frank Knight (1921), op. cit., also Chapter 5 of this book, note 1, than to those of the rational expectations theorists.

(19) Menger's discussion of entrepreneurship is in Menger, op. cit., Chapter 2 of this volume, note 6. See especially pp. 160, 172. (20) Schumpeter's ideas on entrepreneurship were developed in a number of writings beginning from his 1911 book in German, translated by Redvers Opie in 1934 as Joseph A. Schumpeter (1934), *The Theory of Economic Development*, Cambridge, MA: Harvard University Press. A mature statement, which explicitly puts the stress on creativity, is Joseph A. Schumpeter (1947), "The creative response in economic history," *The Journal of Economic History*, **7**(2) (Nov.) 149–59. This is probably the best brief reading for those new to Schumpeter. (21) On the distinction between small businesses and entrepreneurial innovators, see Eric G. Hurst and Benjamin W. Pugsley (2011), "What do small businesses do?" *Brookings Papers on Economic Activity*, **43**(2) (Fall), 73–137. (22) An example of the discussion of intrapreneurship may be found in David Armano (2012), "Move over entrepreneurs, here come the intrapreneurs," *Forbes*, 21 May, available online at http://www.forbes.com/sites/onmarketing/2012/05/21/

move-over-entrepreneurs-here-come-the-intrapreneurs/, as of 23 August 2012. On social entrepreneurship see J. Gregory Dees (2001), "The meaning of 'social entrepreneurship,'" available online at http://www.caseatduke.org/documents/dees_sedef.pdf, also as of 23 August 2012. Dees' very readable paper also provides a good introduction to some of the meanings attached to the word "entrepreneur" in management thinking.

7. Governance

We have seen that people can often benefit from the formation of groups that work together. Using the terminology of cooperative game theory, we call those groups coalitions in general. Coalitions can create value in a number of ways: by coordinating the division of labor and the production and use of higher order goods, including knowledge, which enhance the productivity of labor; by organizing the reallocation of goods, services, and resources in ways that are more productive or better adapted to the different preferences of consumers; and by sharing independent risks so that these risks can be managed more cheaply. If information were free, the activities of these coalitions would always be efficient, at least from the point of view of the members of the coalition. However, as we also have seen, information is not free, and this leads to a number of arrangements by which coalitions and individuals adapt to limited and costly information. Among the most important are money and organizations. An organization is a coalition that persists over time, with some persistent rules of procedure that may include a hierarchy of authority. In this book, the term "governance" refers to the hierarchies of authority and decision-making, formal rules of procedure, and informal social and moral norms that influence the day-to-day actions of a coalition, and particularly its responses to unanticipated problems and opportunities. This chapter is concerned with the governance of organizations.(1)

In the previous chapter a number of examples were given to show how the cost of information explains many of the most familiar aspects of our economic lives, including organization. This chapter will focus instead on just *how* organization economizes information.

WHEN NON-COOPERATIVE DECISIONS ARE COOPERATIVE

An important theme of this book, and especially of Chapter 3, has been the contrast between cooperative and non-cooperative decisions and arrangements and the fact that the economic world we live in is a blend of both kinds of arrangements. In part, this is because information is costly and the rationality of real human beings is bounded, as discussed in Chapter 6. Forming a coalition to carry out a common course of action is a transaction, and if the transaction cost is sufficiently high, it would not be worthwhile forming the coalition. Then again, boundedly rational agents may act non-cooperatively simply because their rationality is bounded. In deciding whether to seek a cooperative agreement or not, the agents will use decision processes that are cheap but also subject to error, and so they may base the decision whether or not to cooperate on an estimate of the benefits that turns out, in practice, to be mistaken – one way or the other.

But there is another reason why cooperative and non-cooperative decisions are mixed, and it will be more important for our purposes in this chapter. This is illustrated by the example of the auction in Chapter 4. We will revisit it briefly. In that example the item to be auctioned was a unique Old Master painting, and each bidder was assumed to know exactly how much she or he would be willing to pay for it. As in the economic and game theoretic literature on auctions, we assumed that bidding decisions are made non-cooperatively. In part this reflects a social convention that bidders in an auction *should* bid non-cooperatively.(2) Nevertheless, as we observed in Chapter 4, it is a cooperative enterprise to organize an auction; and those who participate in the auction, as buyers and sellers, are members of a coalition. All expect to benefit on the average, by having the opportunity to buy or to sell under the rules of the auction. But the rules of the auction demand non-cooperative bidding decisions. Why so?

How would the transfer of the Old Master be arranged in a world of purely cooperative decision-making? If information were free, the seller would simply identify the efficient buyer at no cost, and they would come to an agreement between the two of them. They would have to bargain over the division of the surplus. Each knowing (at no cost) the bargaining power of the other, they would have no difficulty arriving at their bargain.

In a world of costly information, however, it will be necessary to produce information as to the willingness to pay of the potential buyers, and which, therefore, is the efficient buyer; and it will also be necessary to produce information as to the bargaining powers of the buyer and seller. For this purpose we might see a coalition of the seller with all those who *might*, before any information is produced, have the largest willingness to pay. But the production of information on willingness to pay can be quite costly. Potential buyers have an incentive to conceal their willingness to pay, to bluff. Of course, bluffing is non-cooperative – it is a deviation from the common goal of the coalition, a cheat – but there will always be costs of preventing cheating. In addition, each member of the coalition will expect some share of the surplus from the transfer, depending on his or her relative bargaining power. The process for producing information on willingness to pay and on bargaining power in this case is the process we call bargaining, and it will cost some time and effort of the bargainers. The costs of the bargaining process also include some risk that bargaining will break down and the transaction will not be made efficiently or will not be made at all.

Suppose, instead, that the coalition of the seller and the potential buyers choose to decide the transaction by non-cooperative bidding according to the rules of an auction. In that case each person needs know only her own willingness to pay, usually a rather cheap bit of information. Thus, the non-cooperative decision-making requires much less information in this instance. Indeed, it will often be true that non-cooperative decision-making requires far less and cheaper information than cooperative decision-making. Having agreed to settle their common business by means of non-cooperative decision-making within the rules of the auction, the group gain a larger surplus than they might obtain if they were to decide the sale by a strictly cooperative procedure. In part, this works because the rules of the auction lead to an efficient result, that is, the sale of the painting to the buyer whose willingness to pay is greatest.

In the process, however, the auction also determines how the surplus is divided: it is divided in certain proportions between the efficient buyer and the seller, and nothing to anybody else. From that point of view, yet another advantage of the auction is that it makes bargaining unnecessary. So the costs of bargaining are saved. While nobody except the buyer and the seller gets any direct share in the auction, other potential buyers may be willing to participate in the auction nevertheless. Not (initially) knowing who will be the efficient buyer, each participant in the auction has some probability of a benefit, and that is what leads each of them to participate in the auction. In that sense – at the stage of deciding whether to participate in the coalition constituted by attendance at the auction – each potential buyer gains a larger mathematical expectation of a benefit from the auction coalition than he or she would have otherwise, and so the non-cooperative bidding realizes the cooperative objective of a mutually beneficial re-assignment of the ownership of the Old Master. We should also notice, however, that the auction *transforms* the bargaining power in the game. The auction divides the surplus from the coalition between the seller and the efficient buyer – nobody else has any bargaining power whatever.

Let us recapitulate, in a more general way, what the auction example shows. A group of people have an objective that is mutually beneficial to them, so far as they can tell in the light of information that they have available to them at the beginning. In order to attain that objective, they establish rules for a non-cooperative game. The non-cooperative game, in the example the auction, has the property that when it is played non-cooperatively, the outcome we can reasonably expect – in this case a dominant strategy equilibrium – is efficient. Moreover, the information necessary to play this game efficiently is available at little or no cost. Thus, the group attain their cooperative objective by acting non-cooperatively.

RULES OF THE GAME

As we said in Chapter 4, the auction is a very special case. As we have seen, in many games non-cooperative play leads to inefficient results, at least when the results correspond to what game theorists consider rational equilibria for non-cooperative games, such as dominant strategy equilibria or Nash equilibria. That is what led us to investigate cooperative coalitions in the first place. The auction in the example is a special kind of game in that the dominant strategy equilibrium is efficient. Further, in this chapter, we have so far ignored a complication that was mentioned in Chapter 4. Here it is: some of the potential buyers may cheat by forming a (cooperative) bidding ring. If they do, then they are violating the social norm or general agreement that bidding will be non-cooperative. Thus, in practice, the cooperators in the auction will need some assurance that everybody in fact is following the rules – an information cost we did not consider – and there may be a need for some sort of enforcement to punish those who do not bid non-cooperatively. But a game with such rules is a more complex (and information-costly) game. Finally, the auction is special in that the information necessary to play the non-cooperative game is easily obtained. We have assumed that each person knows what the Old Master is worth to him- or herself. If someone were buying the painting to resell it – so that he or she would have to estimate the profit from the resale – that in itself would result in a more complicated game in which, quite possibly, the Nash equilibrium is not efficient.

Suppose, then, that a group of people have some common objective that is in their mutual interest – not necessarily the sale of an Old Master painting, but perhaps a quite different objective – and they want to design a non-cooperative game by which means they can realize that objective with a minimum of informational cost. They will need to design a game that meets certain requirements:

1. It should be incentive-compatible. The rules of the game (which some would call "the social mechanism") are incentive-compatible if the game has a unique non-cooperative equilibrium that corresponds to the cooperative outcome desired.

2. The game should be resistant to attempts by subgroups to cooperate among themselves, to the disadvantage of other members of the coalition, that is, resistant to cooperative deviations. This is where auctions fail when there are bidding rings.
3. Since the game is to be played by boundedly rational human beings, it should be cognitively simple, so that the agents can play the non-cooperative game at little or no informational cost.
4. The rules should not discourage agents from acting cooperatively in the interest of the whole coalition, over and above what the rules require of them.

This probably sounds like a very demanding list of requirements – and it is. Nevertheless we know a good deal about the first two items on the list, incentive compatibility and resistance to cooperative deviations. The research field known as social mechanism design has focused on the design of games that are incentive-compatible, and a good deal has been learned.(3) Nevertheless (as we will see) there are some negative results as well as positive ones. And while most of the research on social mechanism design has made the assumption that decision-makers are perfectly rational, cognitive simplicity has also been taken into account in some of that research, so we know something, at least, about item 3. On criterion 2, the 2012 Nobel Memorial Prize in Economic Science, awarded to Shapley and Roth, honored work on choice procedures that satisfy the criterion, in the case of matching individuals of distinct types to form two-person coalitions.(4) One key result from this research is that our first two criteria can conflict – that there may not be a procedure that is both incentive-compatible and stable against cooperative deviations, in general. However, there is a great deal more to be learned.

In a world of costly information and boundedly rational decision-makers, criterion 3, that the rules should be cognitively simple, is particularly important. For that purpose, we can again use the guidelines suggested by Baumol and Quandt in their paper on optimally imperfect decisions, quoted in the previous chapter

In any case, what the criteria 1–4 describe is an ideal. Incentive-compatible rules (in the first instance) operate by aligning the rewards to individuals with the benefits to the coalition as a whole. However, most rules that influence our decisions in the real world are at best only approximately incentive-compatible. Many of those real-world rules rely on penalties for violation in order to encourage people to follow the rule rather than acting either non-cooperatively or cooperatively to the disadvantage of other members of the coalition. Penalties require information. In order to impose a penalty it is necessary to know who has violated the rule; moreover, since people have incentives to evade the penalties if they can, still more information is necessary to assure that the penalty is in fact imposed. Since information is costly there will be failures on both sides – failure to detect and/or penalize violations, and penalization of persons who have in fact not violated the rules. This informational cost is often called the cost of monitoring compliance with the rules. Here, as usual, we see a trade-off between the cost of information and the benefits of a more thoroughly cooperative activity by the group. Moreover, the design of rules that are incentive-compatible and/or stable against group deviation is itself a very costly process of information production, since it relies on expertise at a very high level and on creativity at the same level of expertise. Thus, the investment of designing incentive-compatible rules may not pay, even where it is feasible.

We should pause to observe that people will sometimes act cooperatively, in the interest of the whole coalition, even at some cost to themselves. They may also carry out penalties against those who act non-cooperatively despite the fact that the penalties make them also worse off. Both everyday experience and some experimental evidence confirm this. That is the reason for criterion 4. A classic failure in this connection was discussed by Titmuss: in some countries, blood for transfusions and blood banks could only be donated voluntarily, with no payment permitted, while in other countries blood could be purchased by blood banks. With payments allowed, some people would donate blood voluntarily, but Titmuss argued that payment would discourage voluntary donations, and might therefore make the shortage of blood worse.[(5)]

A further qualification is that cooperative strategies are far more likely when relationships among the "players in the game" are stable and long-lasting, so that the non-cooperative game is played again and again, repeatedly. This idea was present in game theory from the first.[(6)] Because it was believed to be true for years before it was given mathematical form, it is often called the folk theorem. What the mathematical literature tells us is that when play is repeated *with a high probability* following each step in the repetition, then it is *possible*, depending on the details, that the cooperative strategies will also be a Nash equilibrium. However, it is not a unique Nash equilibrium. In general there will be infinitely many Nash equilibria, including inefficient non-cooperative equilibria as well as the efficient cooperative ones. This negative result needs not discourage us very much. In his 1959 paper,[(6)] Aumann already suggested a way to narrow the field. For any game with two or more Nash equilibria, suppose that a coalition forms among some or all of the "players in the game," and the coalition chooses among the different Nash equilibria of the game. If there is a Nash equilibrium that would not be rejected by any coalition, then it is a *strong* equilibrium. Aumann identified the strong equilibria with cooperative solutions of games that are played repeatedly. Moreover, some experimental evidence suggests that, with boundedly rational players, repeated play will bring about some cooperation even in games that would not be played cooperatively by perfectly rational players. While there is more to be learned, it seems likely that a stable coalition can often realize a cooperative strategy as a strong Nash equilibrium. This in turn could correspond to an implicit contract among the members of the coalition, such as we discussed in the previous chapter.

Nevertheless, a coalition will often find it best to establish rules that are only approximately incentive-compatible, and to rely on rules with informational costs of monitoring, and then to tolerate or encourage non-cooperative decision-making within the rules. Conversely, non-cooperative decision-making, within an appropriate set of rules, can advance cooperative objectives, and violations of the rules, either non-cooperatively or on the basis of a cooperative agreement among the violators, can frustrate the cooperative objectives.

On the other hand, there are some cases in which rather simple rules can solve problems in the coordination of cooperative action, and we would expect that those simple rules would play a large part in many organizations. And indeed they do.

HIERARCHY

One of the most common characteristics of organizations in a modern economy is hierarchy. In some ways this seems to go against the spirit of modern society, as hierarchy has also been central in traditional societies. Hierarchies in modern societies are often described as meritocracies, in which status within the hierarchy is an attained status, by contrast with traditional societies in which status is ascribed, and often hereditary. But we shall see that hierarchy solves a problem that arises in organizations that rely on division of labor or on coordinated action, such as production and military organizations. Thus, we should not be surprised to see a widespread role of hierarchy in modern organizations.

Division of Labor

As we have observed, non-cooperative decision-making *may* be inefficient, but in some games some non-cooperative equilibria may be efficient. However, other difficulties can arise if the non-cooperative equilibrium is not unique. Within this category are games called coordination games.

For this chapter, a coordination game is one in which there are at least two distinct Nash equilibria, both of which are Pareto-efficient.* As we observed in Chapter 3, these games are particularly important when division of labor plays an important role in production. Let us revisit the Division of Labor Game: see Table 3.4 in Chapter 3.

As we saw, there are Nash equilibria when the two workers choose different tasks, and these strategy combinations are Pareto-dominant over every other strategy combination, including the third Nash equilibrium where both workers work alone. Further, rationalization can fail badly in an example like this, since rationalization may not lead to an efficient Nash equilibrium. To achieve the efficient (and cooperative) solution to this game, the two workers must somehow choose one of the two efficient Nash equilibria and base their strategy choices on it. Now, this is a problem but, in practical terms, not a difficult one. With a little cheap information, it can be solved. For example, in a two-person game, the two workers can flip a coin, choosing one equilibrium if the coin lands heads and the other if tails. Another possibility – which will make more sense in a larger game with more participants – is to trust one person, a player in the game or a trusted third party, to make the decision among the different Nash equilibria. As we observed in Chapter 3, in a proprietary enterprise, it could make sense for the proprietor to be the trusted third party.

In another sense, though, trusting one person to make the choice between the two Nash equilibria does not solve the problem at all, but only puts it off to a higher level. In place of the original game we now have a new game in which the strategies determine the choice of the person to choose among the Nash equilibria. And this is another coordination game, since the choice will be made only if the players in the original game agree on the choice of the coordinator. Agreement will be relatively easy in the two-player Division of Labor Game, but in a larger and more complicated game – in which the payoffs for one Nash equilibrium may be relatively more favorable to one player than to others – there may be something to be gained by designating oneself as the coordinator. (The second Division of Labor Game in Table 3.5, Chapter 3, provides a very simple example.) And in a more realistic game it may be that the proprietor can gain at the expense of the workers by choosing one Nash equilibrium rather than

* For the purposes of this chapter I make no distinction between coordination games – in which it is efficient for all players to choose the same strategy – and anti-coordination games, in which it is efficient for the players to choose different strategies. There are some differences in the informational problems posed by the two cases. On the other hand, in a mathematically formal approach to games in normal form, the phrase "players choose the same strategy" is often not very meaningful.

another, so that trusting the proprietor may not come as naturally, either. The Division of Labor Game is intentionally a very simple instance of a game that demands coordination; in less simple instances the problem could be more difficult.

The problem arises with particular urgency in armed conflict. In armed conflict, coordination of actions of attack or defense are a necessary condition for the survival of the unit, and the chances of survival will usually be worse, for every member of the unit, if the coordination of actions fails. At the same time the coordination of attack or defense will often impose much greater risk and hardship on some soldiers than on others, as a condition of success. Moreover, leaders responsible for coordinating the actions of their units will often be removed by hostile action. In those circumstances, few things could be more urgent than the need to determine who is in charge – who will be responsible for coordinating the action of the unit. It will be obvious that rank plays the key role here. The new commander will be the person of highest rank among those present. If two are of equal rank, then it will be the one with greater seniority; if equal seniority, then the older.

Ranking, then, is a general solution to coordination games. The solution is a rule that says that, whenever there is a coordination problem in which two or more persons could decide among the distinct efficient Nash equilibria, the choice is made by the person of higher rank. Moreover, this rule is incentive-compatible in the Division of Labor Game, since the decision by the person of highest rank results in a unique non-cooperative solution. Since it is a Nash equilibrium it is self-enforcing – in the ideal case no penalties for violations are necessary and so no further information cost for those purposes is required. (In practical reality this will often be more complicated and enforcement will be needed, especially in the case of military action and combat.) In addition, the rule of hierarchy fits well with the Baumol and Quandt conditions of cognitive simplicity from the previous chapter. Further, since division of labor is pervasive and crucial in modern production, we would expect to see hierarchies of rank in production organizations, as, of course, we do. These facts probably account for the widespread use of rules of hierarchy in large coalitions of many kinds.

There is one further condition, though. The condition is that the hierarchy has to be generally known and determined *outside the game*, before the game begins. Otherwise, we will once again just put the problem off one more stage, to a competition for rank; and competition for rank tends to be a very inefficient, even destructive, process.(7) This is probably the reason why formal acts of recognition and affirmation of rank, such as military commissions and letters of appointment, play an important role in many organizations to which rank is important. In other cases we may need rules to determine what the ranking will be.

Voting

One such rule to determine the ranking is by voting. Again, recall that in Chapter 3 we remarked that the workers in a work group with a Division of Labor Game might hire someone to do the work of coordination, and in that case they would constitute a worker cooperative. (This is a somewhat different use of the term "cooperative.") In the case of a worker cooperative, the cooperators probably would elect their coordinator by voting. Voting, of course, may also be a way of choosing other kinds of rules besides rules of hierarchy, and plays a role in many sorts of organizations, including both political and non-political ones. Here, though, we are mostly concerned with voting as a way to select a coordinator or to determine rank for that purpose.

Is voting incentive-compatible? The answer is no, but for present purposes, that might not be very important. If we are voting to designate the person who will choose among Nash equilibria in a Division of Labor Game, so that each of those Nash equilibria is efficient, then the problem does not arise. If the Nash equilibria are only approximately efficient, then electing the chooser will not introduce any inefficiency beyond that inherent in the rules themselves. Of course, it may make a great difference to *distribution* of the mutual benefits from working together. In a conventional corporation, in which the only members of the coalition who have voting rights are investors, the Nash equilibrium chosen is likely to be the one most favorable to the investors. In a worker cooperative, in which only employees have voting rights, it is likely to be one that is more favorable to the employees.

Since voting is a key part of many political organizations, we will return to the discussion of voting in the next chapter.

Span of Control

We have seen that hierarchy provides a solution to problems of multiple Nash equilibria when division of labor or assignment of tasks and missions need to be coordinated. For this purpose, there might in principle be just one coordinator – the hierarchy helps to select him or her, but once he or she is selected, the hierarchy can remain in the background. This could work if information were free and the coordinator were an "absolutely rational [being] whose capabilities of reasoning and memorizing are unlimited."(8) Since those things are not so, each coordinator is likely to have a limited span of control.[9] As a consequence we observe a hierarchy of coordinators in any very large organization. This hierarchy is usually created from the top down. A first-level coordinator is chosen by a pre-existing hierarchy, or by a vote of the board of directors or of the members of the coalition, and among the duties of the first-level hierarchy will be the appointment of those at the next lower level, and so on down the chain until a hierarchy is in place. This hierarchy tells each subgroup within the coalition what roles they will play that determine the Nash equilibrium in their particular group.

A further note of realism is wanted here. In most organizations the rules of the game will only be approximately incentive-compatible at best. The intended coordinated activity of the group will be a Nash equilibrium only to the extent that rules and policies are enforced by penalties for those who violate them and rewards for those who comply. The coordinators thus find themselves also the judges of who has complied and the assessors of penalties and rewards. This, of course, is called supervision, and so the hierarchy of coordinators becomes a hierarchy of supervisors. This increases the cognitive work of those in the hierarchy and so narrows the span of control for each of them, making for taller and narrower hierarchies. (Some students of management have believed that improving communication technology would make for a larger span of control and thus flatter hierarchies.) However, a key point for this section is that hierarchies would be expected even if there were no need for supervision and control, given the importance of the division of labor in modern production organizations, and so the likelihood of plural Nash equilibria.

Interim Summary

Whenever division of labor or complementary tasks or missions are important, there may be two or more Nash equilibria, which creates a danger that rationalization might lead to inefficient, non-Nash outcomes. One solution to this problem is a hierarchy, and for that reason, we would expect production organizations to have a hierarchical structure, as indeed they do. Given that most organizations rely on rules that include penalties and rewards to reconcile the objective of the group with Nash equilibrium, this hierarchy of coordinators is also, as a rule, a hierarchy of supervisors as well. Given the limits of human information processing and the cost of information, each coordinator is likely to have a limited span of control, and this reinforces the need for a hierarchy and makes for a taller and narrower hierarchy than might otherwise be observed.

GOVERNANCE OF CORPORATIONS, IN PARTICULAR

Adam Smith was quite opposed to corporations, or, as they were called in his time, joint stock companies. He did not believe that they could be competitive except in very special circumstances. He wrote:[(10)]

> The directors of such companies, however, being the managers rather of other people's money than of their own, it cannot well be expected that they should watch over it with the same anxious vigilance with which the partners in a private copartnery frequently watch over their own. . . . Negligence and profusion, therefore, must always prevail, more or less, in the management of the affairs of such a company.

By the time that John Stuart Mill was writing on economics, however, corporations were more common and it was clear that they could have a competitive advantage in at least some important economic functions, such as building railways. Accordingly, Mill was more optimistic, writing that:[(11)]

> it is not a necessary consequence of joint stock management, that the persons employed, whether in superior or in subordinate offices, should be paid wholly by fixed salaries. There are modes of connecting more or less intimately the interest of the employees with the pecuniary success of the concern . . . it is a common enough practice to connect their pecuniary interest with the interest of their employers, by giving them part of their remuneration in the form of a percentage on the profits.

In these passages, Smith set out what came to be seen in the twentieth century as the central problem of corporate governance, and Mill provided the basis for the widely accepted solution.

A corporation may be thought of as a coalition of shareholders, or more correctly as a coalition of shareholders, employees, and customers for production and exchange in which the shareholders have the sole voting power in selecting the management. The former interpretation is the more orthodox though, and thus we will follow it here. The difference in interpretation will make no difference to the discussion in any case. The problem, then, is to obtain the management on terms that are incentive-compatible in that they align the incentives of the management with those of the coalition.

By the mid-twentieth century, many economists had despaired of a solution, and the view was widespread that corporations were managed according to the discretion of the managers. There were several hypotheses as to how this managerial discretion might predictably be used – for example, that managers might be more interested in increasing the scale of the organization or its rate of growth than in maximizing the shareholders' profits.(12) In the latter half of the twentieth century, however, Mill's view of things was revived. New emphasis was put on ways to compensate managers so that their interest would be aligned with that of the shareholders, such as compensation with options to purchase shares. This turning in economists' thinking was signaled, and led, by a paper of Jensen and Meckling.(13)

Formally, the directors are elected by the shareholders, and we might hope that this would be enough to align the directors' interest with that of the shareholders, so that the directors in turn would either supervise the management strictly, or establish terms of compensation for the management that would be incentive-compatible to a considerable degree. However, this overlooks some things that Smith and Mill do not overlook. First, remember, voting is itself a costly cooperative act. Thus, most shareholders, motivated by personal gain, do not vote. (I was once advised: if you don't like the management, don't vote for new management – sell the stock.) Smith was aware of this, writing that "the greater part of those proprietors [the shareholders] seldom pretend to understand anything of the business of the company, . . . give themselves no trouble about it." And Mill had reservations about the motivations of the directors:

> Even the . . . board of directors, who are supposed to superintend the management, and who do really appoint and remove the managers, have no pecuniary interest in the good working of the concern beyond the shares they individually hold, which are always a very small part of the capital of the association, and in general but a small part of the fortunes of the directors themselves.

Where there are one or more very large shareholders, those particular shareholders may have reason enough to take an interest in the details of managing the company so that their vote, and service on the board of directors, is a profitable non-cooperative course of action. On the other hand, a key requirement of corporation laws in general is that the benefits from the company are distributed in proportion to share ownership, whether in the form of dividends or of policies that increase the market price of the shares. Thus, if the large shareholders see to it that the company is efficiently managed, the smaller shareholders can rely on that and need take no interest in the management of the firm whose shares they own. (In the language of economics, this is a positive externality from the large shareholders to the small shareholders.) This would lead us to expect that the directors and the management would have more discretion to pursue their own interests in corporations with no large shareholders.

Most of the discussion of corporate governance has relied on non-cooperative game theory and economics, and many of the mathematical models are simplified by assuming that there are only two or three persons in the game. Typically, they are principal–agent models. The owners are considered to be the principal, and the executive or manager of the company the agent. The only cooperative elements in most such models are bilateral contracts, and even in these contracts usually take the form of a (non-cooperative) offer by the principal, with the agent (non-cooperatively) accepting if the offer is no worse than his or her alternative. Bargaining power and other cooperative phenomena are assumed away by such ad hoc specifications of the model. The objective of the principal is to determine contract terms that are incentive-compatible in the sense that the agent, acting non-cooperatively, makes decisions that advance the interest of the principal.[(14)] In the spirit of this chapter, we might simply identify the coalition as a whole as the principal, and see corporate governance as a particular case of a coalition's attempt to save information costs by establishing rules within which non-cooperative decisions may be relied on. But there is a difficulty here.

In particular, models of corporate governance do not clearly state whether the owners, as principal, are shareholders, directors, or both. They do, however, assume that the principal is a unitary decision-maker. But the fact that there are many shareholders is the root of the problem. By the assumption that the owner is a single decision-maker, collusion by (some) executives and directors is assumed away. Put otherwise, criterion 2 from the first section above is ignored. There seems very little research on this topic, and the little that exists is in the spirit of the previous literature on corporate governance.[(15)] That is, it attempts to find the terms for a contract to align the rewards of directors and executives with those of the shareholders in such a way that collusion is not profitable or, if it occurs, promotes the interest of shareholders. However, this overlooks a key point. It is this: the directors are the elected representatives of the shareholders, so the terms of the contract between the corporation and its executives *and its directors* are chosen by the directors themselves. If they have incentives to choose an efficient contract, then they will not need an efficient contract to establish the incentives, and if they do not, then presumably the efficient contract form will not be chosen.

If the problem is that the shareholders do not vote for directors who will advance their interests, because it is not non-cooperatively rational for them to do so, then (as we saw before) the directors and the executives will share a great deal of discretion. This will include the discretion to collude against the interest of shareholders. Perhaps this explains the seeming tendency – much discussed in the popular press – for executives to receive "incentive" bonuses even when there are no profits and the corporation has been unsuccessful.

In this section until now, the corporation has been treated as a coalition of shareholders. While this is conventional it is not conclusive. Employment and exchange are cooperative relations. Thus, a corporation might be better thought of as a coalition for production and exchange, with employees and customers, along with shareholders, as members.[16] Of course they are not members on the same terms. The limitation of voting rights to shareholders, and their nominal entitlement to the residual generated by the firm, give them an advantage in bargaining power (although their difficulty in controlling the corporate executives who are their agents may offset that somewhat). Nevertheless, employees and customers will have *some* bargaining power in the coalition.

In a world of free information, the coalition for production and exchange would first maximize the surplus generated by the coalition and would then distribute it according to bargaining power. Since there are information costs, it will be costly to recruit new employees and customers, and each such recruitment will generate some incremental surplus, even in highly competitive conditions. Since it remains the objective of the coalition to maximize the surplus (rather than the profits, the residual) much of what has been said about corporate governance needs not be modified. The cost of recruiting new members implies that monopoly or monopolistic competition will be the usual case. However, some second-degree price discrimination is to be expected, as a means of attaining the efficient level of production (thus approximately maximizing the surplus) while distributing the surplus according to bargaining power. It follows that monopoly will not distort the allocation of resources, as in conventional microeconomics, though it will shift the balance of bargaining power. The same will be true of monopsony power in labor markets.

AN ECOLOGY OF ORGANIZATIONS

We have seen that cooperative coalitions in our world of costly information will often have certain characteristics, such as hierarchies and rules with rewards for compliance and penalties for deviation, and within those frameworks, tolerate or even encourage non-cooperative decision-making within the group. We might imagine a world in which all organizations face similar challenges and so adopt a single common form. However, the world we live in does not seem to be that kind of world. Instead, we observe a wide range of organizational forms.

Uniqueness

It seems clear that the optimally imperfect rules will vary from one sort of organization to another. A military unit, as we have seen and as experience teaches, will rely strongly on a hierarchy of command and punishment of those who deviate. By contrast, an artist's cooperative may have nothing in the way of hierarchy. The rules of one such cooperative require the members to display some of their art work for sale, specify the mark-up that goes to cover the costs of the cooperative, and require the members other than the manager to contribute some hours minding the store. The one member with some knowledge of accounting is routinely elected manager and serves gratis. Most organizations will be between these two extremes.

But – even in given circumstances, for a particular coalition – it is unlikely that there is any unique set of optimally imperfect rules. A set of rules that relies strongly on individual initiative and rewards will generate one kind of costs, while a set of rules that relies more strongly on hierarchy and punishment will generate another kind, but depending on the details, the two kinds of costs may amount to the same. We recall criterion 4: the rules should not discourage agents from acting cooperatively in the interest of the whole coalition, over and above what the rules require of them.

Thus, even if all organizations faced the same challenges, we might observe a range of organizational forms. But, in fact, organizations also occur in different forms because the pressures on them are different. Organizations, like animals, are born and die and the proportions of different forms in the economy as a whole will reflect the conditions of their initiation and conclusions. Private proprietary businesses seem both to have high rates of formation and closure, and corporations somewhat less. For other forms of organization still other conditions will prevail.

The Social Economy

For-profit enterprises play a key role in most modern economies. This chapter has argued that these organizations are best understood as imperfectly cooperative coalitions of (at least) owners and employees (and, more correctly, also of their customers). This class of enterprises is distinguished by the fact that a proprietor or group of proprietors are the de jure owners of the residual from the activities of the group and they (or their representatives) have the sole voting power in determining the hierarchy of the group. Important as these organizations are, they are not the only ones that engage in production and in sale of the product. State enterprises, of course, remain important in many countries. In the later twentieth century, the distinction of state enterprises from for-profit corporations has become somewhat blurred as (on the one hand) sovereign wealth funds, especially in petroleum-exporting countries, have purchased the shares of corporations and (on the other hand) semi-privatization of state enterprises has often taken the form of the sale of shares representing a proportion of the ownership of the enterprises. Little more needs be said here about state enterprises. However, this subsection will briefly discuss three other enterprise forms that, in Europe, are often said to constitute the social economy (which usually also includes state enterprises).

Non-profits (NGOs)

In the United States, most if not all of the 50 states have statutes permitting the establishment of what are commonly called non-profit corporations. Strictly speaking, these organizations are not forbidden from earning profits. (Tax treatment as a non-profit under US federal tax law may set some effective limits on profitability.) Rather, they are forbidden from distributing profits to owners, or to employees or to any other stakeholders. Rather, the non-profit corporation is expected to devote whatever resources it may have, whether derived from profits or from other sources, to its mission. Laws vary from state to state, and vary widely in other countries as well, but broadly speaking, the existence of the non-profit corporation derives from its charter, and the charter states the mission. Missions range from the distribution of charitable gifts to education and medical care to the free provision of literature that supports free markets, and widely in many other directions. The corporation is overseen by a board of trustees who recruit the executives and also select their own successors and who bear some legal responsibility to assure that the corporation does indeed act to advance its mission. Outside the United States these are more commonly known as non-governmental organizations, NGOs, since many serve purposes (such as education and medical care) that historically have often been the missions of government organizations.

The non-profit sector of the American economy is often described as the fastest-growing one; that is, growing faster than either private enterprise or government organizations. It is clear that there is an ecological-economic niche in which they can prosper. Why? Many non-profits are originated by founding endowment gifts, or rely on gifts from time to time to continue their activities. Hansmann[1] argued, along lines very much parallel to Williamson's ideas, that non-profit enterprise could have an advantage in activities supported by gifts, since the non-profit status would protect the donors against opportunism by non-profit managers who might otherwise divert the donated resources to purposes other than those intended by the donors. Put otherwise, the prohibition on distribution of the non-profit's profits or other assets makes it possible for donors to trust that their intentions will be carried out.

Many non-profits in fields such as health and education charge for their services and, while they may also solicit gifts, do not rely primarily on gifts for their routine operation. However, their clients, students, and patients may choose non-profit suppliers for reasons that more or less parallel the motives of donors. In these fields the quality of service is both crucial to the benefits obtained by the client and very difficult to specify by contract. In the educational context, for example, non-profits may offer a wider range of programs, some of which do not cover their cost. This would provide undecided students with opportunities that most may choose not to exercise but nevertheless value at the time of their enrollment. The value set on this sort of freedom of choice is known in the economics literature as an option demand,[17] and, like a public good, provision for option demands can generate positive benefits although it cannot cover its costs. Again, we seem to find non-profits where it is important for the client to trust that the mission of the organization will guide its decisions.

In the terms of this chapter, a non-profit corporation should probably be thought of as a coalition among donors, employees, and beneficiaries. We need not reconsider the role of trust as a reason for the prohibition of distribution of profits. So far as governance is concerned, probably only a little needs to be added to what has been said about governance of corporations. However, the difficulty of preventing collusion among the directors or trustees at the expense of others will be even more extreme, in some cases. Essentially the directors of a non-profit are a self-coopting group. Such a group may remain highly committed to its mission, in many cases, but when the directors deviate from the mission of the corporation, it can be very difficult to remove them short of bankruptcy. This is a kind of informational cost, and it can be a crucial one. This will be less of a difficulty where the non-profit corporation depends on ongoing, annual gifts, and more likely to be a difficulty where (as in educational and health institutions) the corporation commands a revenue stream. We observe that many non-profits that do command ongoing revenue streams nevertheless continually solicit gifts, and this may be thought of as a source of information on the commitment and competence of the trustees and as a signal to clients that the non-profit remains a trustworthy coalition.

It does also seem that non-profit enterprises are more likely to be long-term survivors than profit-seeking enterprises. This should not be surprising, since non-profits will take fewer risks. On the one hand, should the risks pay off, the extra profit will not benefit the decision-maker, and on the other hand, should the risks not pay off, the mission may itself be put in danger. Thus, non-profits are less likely to be "born," since initiation of any enterprise is risky, but they also have a lower death rate than profit-seeking enterprises. This is a factor in their relative growth in numbers.

Cooperative enterprises

A cooperative enterprise may be defined as a membership organization that (on the one hand) owns and operates a business of some kind and (on the other hand) in which membership is open to all those who stand in some stakeholder relation to the business, other than as owners. Thus, for example, membership in the coop may be open to customers in the case of consumer cooperatives, to depositors in the case of mutual banks, to farmers who use the services of the cooperative business in the case of agricultural cooperatives, and to employees in the case of worker cooperatives. The managers of the cooperative enterprise then are responsible to the members via majority vote, with one vote per member. Profits are distributed to the members, often as a rebate on their payment for coop-supplied goods and services.

It will be clear that the word "cooperative" is being used in a very different sense here than the sense in the context of "cooperative games" or "cooperative coalitions." The details may be found at the website of the International Cooperative Alliance.[18] On the other hand, certainly a cooperative enterprise is a cooperative coalition, and specifically one in which the coalition's hierarchy and rules are determined by voting and the voting rights limited to the members of the organization. It would seem that this limitation would have some impact on bargaining power within the coalition, shifting it toward those eligible to be members, and this may be a major purpose of such organizations.

The cooperative identity of the enterprise may have an influence on the way in which it is organized. Thus, there is evidence that worker cooperatives have flatter hierarchies than for-profit firms, and spend less on supervision, but nevertheless often attain higher labor productivity. They also commonly use less invested capital per worker, perhaps because they often find it more difficult to raise capital. The Mondragon Cooperative Corporation, probably the most successful and most studied of worker cooperatives, is possibly an exception on this last point because of its distinctive structure, which includes a bank operated in the interest of its member cooperatives.[19]

Farmer cooperatives and cooperative or mutual financial organizations play a very important role in many relatively advanced countries, although their importance varies from country to country. The same can be said of consumer cooperatives, which originated the supermarket in some European countries; in this case, imitative competition from for-profit companies seems to have offset their advantage.

There is reason to believe that foundation of cooperatives is uncommon by comparison with that of for-profits enterprises, while their longevity, in given circumstances, is greater. The rarity of foundations of cooperatives probably reflects two kinds of information costs. First, a number of people need to be brought together to begin a cooperative, so the information required to bring them together is probably a good deal greater than that required to start a for-profit. At the same time, while the information costs of forming a for-profit are largely borne by the proprietor (although some also are borne by his or her employees), the proprietor can also seize a high proportion of the benefits. In the case of a cooperative, both the informational costs and the benefits are borne more broadly by the members. In the absence of an idealistic commitment on the part of the founders of a coop, the motivation to form a coop, net of informational costs, probably is less.

The other informational cost of foundation of cooperative enterprises arises from the fact that many people are simply ignorant of cooperatives. The possibility and viability of cooperative organization is an unknown unknown to them.

Quangos

The term quango may be an unfamiliar term for a rather familiar kind of organization. It is from the acronym for quasi-autonomous non-governmental organization. These are organizations that function routinely as non-profit corporations (non-governmental organizations), with missions and charters and boards of trustees or governors who hire their top management and otherwise direct the organization of the company. However, they may be founded by government action and they may receive government subsidies, while, on the other hand, government may appoint some or all of the directors. Beyond a specific subsidy, quangos are often responsible for their own costs and commonly charge something for their services. Thus, they occupy something of a halfway house between the public and private sectors. Among examples in the United States are the United States Postal Service and Temple University and the University of Pittsburgh in the state of Pennsylvania. The United States Federal Reserve System has much in common with a quango although it is legally constituted as a cooperative of bankers. There are a number of important quangos in Britain, which seems to be the source of the name.

We might think of a quango as a coalition among employees and beneficiaries, as well as any private sector donors. Should we consider the government as a member of the coalition? We shall leave this question unresolved. Nevertheless, government, like other donors, may treat the prohibition on distribution of profits as a guarantee that the quango will make the intended use of the donations it receives. In most ways, the governance of a quango will parallel that of a non-profit corporation without government connections. An exception is that the appointment of some or all members of the board of trustees by government may reduce the informational cost of removal of incompetent or opportunistic trustees. However, government has its own information costs. One symptom of these costs is that the objectives of the government may shift suddenly, and this instability can create crises for quangos that non-profits without government connection would not face.

CHAPTER SUMMARY

In previous chapters we have observed that there are many ways that people may obtain mutual benefits by adopting a common course of action, but also that there are some obstacles to realizing this potentiality. In the previous chapter the cost of information was pointed out as an important obstacle, and many examples were given of how the cost of information explains familiar aspects of our economy. This chapter has focused on organization in particular, and argued (1) that the informational costs of cooperative decisions are often greater than those of non-cooperative decisions, so that (2) one way for a cooperative coalition to proceed is to select rules for a non-cooperative game and rely on the members to act non-cooperatively within those rules. (3) Ideally such rules would be incentive-compatible, that is, such that non-cooperative decisions would be efficient. (4) Nevertheless, non-cooperative rules will also influence the distribution of the benefits, and perhaps not in ways that the cooperative group would wish; that is, in ways that shift bargaining power. This may set limits on the rules the group would wish to choose. (5) Moreover, the potential for cooperative deviations, that is, collusion among some members of the coalition at the expense of the others, will also set limits on the rules the coalition may adopt. With the cost of information per se, this implies that the action of the coalition will only be an imperfect approximation to the cooperative solution that we would observe in a world of zero information costs. One simple rule that can resolve many problems is hierarchy, but, of course, hierarchy also has its costs. The hierarchy may be influenced by voting, and voting provides another set of rules for imperfect cooperation. Finally, the chapter has reviewed some of the discussion of corporate governance, and of other forms of organizations, in the light of the foregoing arguments.

SOURCES AND READING

(1) This chapter follows a trail that has previously been followed especially in the Nobel Laureate work of Oliver Williamson; and Williamson's approach, like the approach in this book, is informal. It is only proper, then, to try to make explicit the relation of this chapter in particular to Williamson's ideas. Some important differences may also be mentioned. See especially Oliver E. Williamson (1975), op. cit., Chapter 6 of this volume, note 2; Oliver Williamson (1986), *Economic Organization*, New York: New York University Press. Like this book, Williamson assumes that contracts are incomplete, rationality is bounded, and information is costly. His focus is on the contract, and he assumes that the parties to the contracts are non-cooperative decision-makers, and, in the case of proprietors, profit-maximizers. This seems an odd mixture of cooperative and non-cooperative elements, since contracts are formed for mutual benefit, that is, cooperatively. For Williamson, an important determinant of the form of a contract is the threat of opportunism, and the need to limit or avoid opportunism. Again, opportunism is non-cooperative action, and its restraint is a necessary condition for cooperative arrangements. Nevertheless, Williamson makes little or no use of ideas from game theory, either non-cooperative or cooperative. This creates some difficulty for his work. For example, Oliver E. Williamson (1975, footnote p. 234) defines opportunism as "self-interest seeking with guile." This is not very satisfactory. A critic such as Edwin West (1987), "Non-profit versus profit firms in the performing arts," *Journal of Cultural Economics*, **11**(2), 37–48, might assert that this is a distinction without a difference, that all self-seeking behavior is opportunistic. In fact West's criticism was directed to a paper by Henry Hansmann (1981), "Non-profit enterprise in the performing arts," *Bell Journal of Economics*, **12**(2), 341–61, which applied opportunism in Williamson's sense to explain non-profit enterprise. Here, again, the issues are clearer in a context of non-cooperative game theory. Opportunism can clearly be understood in terms of subgame perfect Nash equilibria in non-cooperative games in extensive form and Hansmann's explanation of non-profits is put in those terms in Chapter 13 of my book, Roger A. McCain (2014), op. cit., Chapter 1 of this volume, note 8. Further, Williamson consistently relies on bilateral contracts as the sole expression of cooperative action, and seeks to explain governance in those terms as an alternative to market organization, which he treats as given. If it seems presumptuous to criticize a Nobel Laureate in such terms, it should be added that these shortcomings or more serious ones are to be found in almost all of the non-game theoretic literature that has addressed the same range of issues, and that Williamson's work is a clear advance on

the understanding of economic organization in the mainstream economics of the twentieth century. (2) See Chapter 4 of this volume, notes 12 and 13 for a discussion of the literature on auctions.

(3) The 2007 Nobel Memorial Prize honored the work of Leonid Hurwicz, Eric Maskin, and Roger Myerson in social mechanism design, and one can do no better than to refer to the Nobel website for further information, http://www.nobelprize.org/nobel_prizes/economic-sciences/laureates/2007, as of 30 July 2013. Their work is also a good example of the importance of incentive compatibility in the research on mechanism design. On that criterion and the others listed, compare Dilip Mookerjee (2006), "Decentralization, hierarchies, and incentives: A mechanism design approach," *Journal of Economic Literature*, **44**(2) (June), 367–90. This is a valuable survey of a selection of the research findings relevant to this subsection. For a sample of mechanism design that assumes bounded rationality see Vincent Crawford, Tamar Kugler, Zvika Neeman and Ady Pauzner (2009), "Behaviorally optimal auction design: Examples and observations," *Journal of the European Economic Association*, **7**(2–3), 377–87, available at http://www.uclouvain.be/cps/ucl/doc/core/documents/Crawford-2a.pdf, as of 19 May 2013. (4) On the work of Shapley and Roth on mechanisms that are stable under cooperative deviations, again, the Nobel website for the prize awarded to Shapley and Roth is the best beginning point for further information: http://www.nobelprize.org/nobel_prizes/economic-sciences/laureates/2012, as of 30 July 2013. (5) Richard Titmuss (1971), *The Gift Relationship: From Human Blood to Social Policy*, New York: Pantheon Books. Titmuss's reasoning seems consistent with a growing literature in experimental game theory that modifies the self-interest assumption by also including motives of reciprocity. See, for example, Ernst Fehr and Simon Gaechter (2000), "Fairness and retaliation: The economics of reciprocity," *Journal of Economic Perspectives*, **14**(3), 159–81; Joseph Henrich et al. (2005), "'Economic man' in cross-cultural perspective: Behavioral experiments in 15 small-scale societies," *Behavioral and Brain Sciences*, **28**(6) (Dec.), 795–815.

(6) There is a hint in John von Neumann and Oskar Morgenstern (2004), *The Theory of Games and Economic Behavior*, Sixtieth Anniversary Edition, Princeton, NJ: Princeton University Press, that repeated play could lead to cooperation, and Robert Aumann sketched some important details in a 1959 paper, Robert J. Aumann (1959), "Acceptable points in general cooperative n-person games," in Albert W. Tucker and Robert D. Luce (eds), *Contributions to the Theory of Games, Vol. IV, Annals of Mathematics Studies, No. 40*, Princeton, NJ: Princeton University Press, pp. 287–324. However, the large literature on repeated play in non-cooperative game theory stems from the late 1970s. Aumann's Nobel

address provides a good overview: Robert J. Aumann (2005), "War and peace," Nobel Prize Lecture, The Sveriges Riksbank Prize in Economic Sciences in Memory of Alfred Nobel 2005, available at http://nobelprize.org/nobel_prizes/economics/laureates/2005/aumann-lecture.html, as of 9 June 2007.

(7) See Robert H. Frank and Philip J. Cook (1995), *The Winner-Take-All Society*, New York: The Free Press on competition for rank. (8) The quotation is repeated from Reinhard Selten, op. cit., Chapter 1 of this volume, note 9. (9) The limited span of managerial control is an old idea in management, though not old hat – research on the span of control of a manager continues. Economists have also contributed to this research. For an introductory discussion see http://www.economist.com/node/14301444, last accessed 15 November 2013; for a sample of recent research – which is a bit skeptical of the idea that the limited span of control in itself leads to hierarchy – see Oriana Bandiera et al. (2011), "Span of control or span of activity?", Working Paper No. 12-053, Harvard University, accessed 31 October 2012 at http://www.hbs.edu/faculty/Publication%20Files/12-053.pdf, and references there.

(10) The quotations from Adam Smith are from op. cit., Chapter 1 of this volume, note 7 – Book V, Chapter I. (11) The quotation from Mill is from op. cit., Chapter 2 of this volume, note 2 – Book I, Chapter IX, section 2. (12) Williamson discusses this literature at op. cit. (1986, pp. 7–8), and provides a good overview. (13) Jensen and Meckling focus on agency costs, which are among the informational costs stressed in Chapter 6 of this book. See Michael C. Jensen and William Meckling (2000), "Theory of the firm: Managerial behavior, agency costs, and ownership structure," in Michael C. Jensen (ed.), *A Theory of the Firm*, Cambridge, MA: Harvard University Press, pp. 83–135. Strictly speaking, they assume that the problems posed by these costs are optimally solved, and discuss the forms we might expect as a result (p. 87). They argue that rational expectations in capital markets will deter the simplest kinds of managerial opportunism (p. 94). Their comments on incentive compensation systems are candidly described as a conjecture (pp. 129–31). In his 2000 book, *A Theory of the Firm*, Jensen places all this in the context of comparative institutional analysis, although in the 1970s Jensen and Meckling used the term "property rights theory." These terms are associated with literatures that have attempted, at different times, to incorporate transaction costs into a reframed economics based exclusively on non-cooperative decision-making. (14) On the principal–agent approach to corporate governance see, for example, Jean Tirole (2001), "Corporate governance," *Econometrica*, **69**(1) (Jan.), 1–35. (15) A rare exception that does address the incentives to directors is Sylvain Bourjade and

Laurent Germaine (2012), "Collusion in boards of directors," presented at the Congrès AFSE 2012, Université Panthéon-Assas. While this paper makes a strong case that some recent financial problems can be traced to collusion of the directors with the executives, they apply the principal–supervisor–agent three-person game model, which can be traced to Jean Tirole (1986), "Hierarchies and bureaucracies: On the role of collusion in organizations," *Journal of Law, Economics and Organizations*, **2**(2) (Autumn), 181–214. Thus, the shareholders are the principal, the board the supervisors, and the executives the agents. That literature is helpful when there is a single principal, and can be extended in various ways: see, for example, Jean-Jacques Laffont and Jean-Charles Rochet (1997), "Collusion in organizations," *Scandinavian Journal of Economics*, **99**(4) (Dec.), 485–95; Kourouche Vafai (2009), "Opportunism in organizations," *Journal of Law, Economics and Organization*, **26**(1), 158–81.

(16) The idea that a business firm is essentially a *cooperative* coalition of owners and employees is not new and can be found in Masahiko Aoki (1980), "A model of the firm as a stockholder–employee cooperative game," *American Economic Review*, **70**(4) (Sept.), 600–610 and Roger A. McCain (1980), "A theory of codetermination," *Zeitschift für Nationalokonomie*, **40**(12), 65–90. A working paper by Robert Hall extends this to the customers as well, though his model, like so much recent economic theory, is an ad hoc mixture of cooperative and noncooperative decision-making: Robert E. Hall (2008), "General equilibrium with customer relationships: A dynamic analysis of rent-seeking," Hoover Institution, Stanford University, available at http://www.economicdynamics.org/meetpapers/2008/paper_312.pdf, as of 21 August 2011. This perspective is extended in Chapters 8–9 of Roger A. McCain (2013), *Value Solutions in Cooperative Games*, Singapore and Hackensack, NJ: World Scientific.

(17) On option demand see Burton A. Weisbrod (1964), "Collective consumption services of individual consumption goods," *Quarterly Journal of Economics*, **78**(3), 471–7. (18) On the meaning of cooperation in the sense of the cooperative movement, see International Cooperative Alliance (1995), op. cit., Chapter 3 of this volume, note 11. The literature on worker cooperatives is vast, but the flavor can be obtained from some of the very few such studies that have been published in "top" economics journals: Ben Craig and John Pencavel (1992), "The behavior of worker cooperatives: The plywood companies of the Pacific Northwest," *American Economic Review*, **82**(5) (Dec.), 1083–105; (1994), "The empirical performance of orthodox models of the firm: Conventional firms and worker cooperatives," *Journal of Political Economy*, **104**(4) (Aug.), 718–44. (19) See note 12 in Chapter 3 of this volume.

8. A grand coalition of the whole society

Let us suppose that society as a whole were to constitute a cooperative coalition. For the present we will not ask how such a thing might come about, but approach it purely in a spirit of "what if?" In Chapter 10 we will revisit the question of how and whether such a comprehensive grand coalition might exist, but for now it will remain purely hypothetical. Clearly such a coalition would be a very large one (even in a country with a relatively small population). In a world of free information, we might suppose that every individual could be consulted on each and every decision, and unanimity rule required, with bargaining over differences of opinion and interest. In practice it is clear that the informational costs of such a cooperative solution would be impossibly great. Some organization will be required, and that organization might be very complex. A somewhat more possible ideal would be to have some basic decisions made by a trusted third party or a group of trusted third parties. In somewhat that spirit, some early writers on economics and politics (including Machiavelli) addressed their ideas to a "Prince" who would have the power to put them into practice. The next two sections will take the point of view of an executive committee of this hypothetical grand coalition of the whole society.(1)

A MARKET ECONOMY

The executive committee of this grand coalition might consider a command economy as a natural counterpart to a cooperative coalition of the whole society. A command economy is an economy organized from a single center, in which a single decision-maker at the center directs the strategies of all other agents. In a world of free information and perfectly rational agents, the commands would direct decisions that would be efficient and would distribute the benefits of the grand coalition among the members of society in proportion to their bargaining power. If all have equal bargaining power, another ideal case, then the benefits would be distributed equally. But in a world of costly information and boundedly rational human beings, a coalition of this size would be organized as a hierarchy, as we have seen. The commands issued by the center would not be detailed, and the details would be filled in, with coordination of tasks and refinement of heuristic rules successively at each lower level of the hierarchy. (This is, roughly, a brief description of the Soviet system from about 1930 to 1989.) To bring about approximately efficient production and a predictable distribution would require a great deal of costly information, at best.

Therefore, the grand coalition of society as a whole might follow the lines suggested in the first section of the previous chapter: define the rules of a non-cooperative game with a Nash equilibrium that corresponds to the intended cooperative solution, relying on the idea that non-cooperative equilibria require much less costly information. The rules of this game might include rights of property and enforcement of contracts, and similar rules and policies that encourage a competitive market economy. If the world is as neoclassical economics envisions it, the Nash equilibrium of the market game (which we call the market equilibrium) might correspond to an efficient allocation of resources, as in the example of the auction. At worst, it would deviate from this efficient allocation only in predictable ways, as we will discuss in the second and third subsections of this section and in the next section of the chapter.

However, the world "as it is envisioned by neoclassical economics" is a world of free information and unboundedly rational decision-makers, and as we have seen, a command economy would be equally possible in that world – there would be nothing to gain by creating a non-cooperative market game as a way of directing the economy. However, if the informational costs of the non-cooperative decisions are quite small, we may hope that the non-cooperative equilibrium approximates the efficient allocation closely, so that indeed the grand coalition of society as a whole is better off with the non-cooperative market equilibrium than it would be with a fully cooperative solution. The neoclassical welfare economics of the mid-twentieth century assumed, in effect, that the costs of (non-governmental) cooperative decision-making in large groups would be infinite, while that in small groups such as business firms and in groups comprising buyers and sellers in exchange relations would be zero. If that is a good enough approximation, then that kind of economics might be a good guide to making heuristic rules for the grand coalition of the whole society.

Distribution and the "Big Trade-off"

As in the auction, however, the market system would determine the distribution of benefits from the cooperative action of the grand coalition of society, and this might not be the distribution that would be chosen by a cooperative coalition. So, even if the market equilibrium is efficient, the distributional consequences might not be ideal. Some further rules to adjust the distribution of benefits would be needed to approximate the cooperative solution for the society as a whole. In a world of costless information, these adjustments could be made by lump sum taxes and transfers of money. This, of course, is just what neoclassical welfare economics said in the mid-twentieth century. Nobel Laureate Paul Anthony Samuelson, in particular, made that idea central to his version of welfare economics.(2)

However, redistribution through lump sum transfers would require some costly information, and more than that, creates information costs. Redistribution is only needed if the cooperative distribution of benefits differs from the market distribution. As a first step, then, we would need to determine who gains more from the market distribution than the cooperative distribution would give them, and how much; and who does worse in the market distribution than the cooperative one, and how much. This would be a costly bit of information in itself. Much of this information would have to be gathered from the potential gainers and losers themselves. Moreover, within the non-cooperative game of markets and politics, the potential gainers and losers would have no incentive to give accurate information, but quite the contrary. Opportunistic citizens would say whatever they need say to make it seem that they are the losers in the market distribution, to exaggerate the subsidy due to them or to minimize the tax they must pay. Those who would be giving up some of their market gains would also have incentives to resist the loss, so that rules and penalties (known as tax laws) and monitoring and enforcement would be required, creating still more informational costs.

However, if we adopt rules that are not fully efficient, the informational cost may be very much reduced. For example, if taxes are based on income or consumption or some other observable dimension, then the informational cost of administering those taxes could be much less. However, these taxes are inefficient in known ways.(3) Similarly, transfers of benefits to citizens on the basis of observable facts, such as the fact that they are farmers who raise certain crops, will impose less informational costs than the lump sum transfers, although the transfers based on observable facts will be inefficient.

Farmers provide a good example. During the twentieth century, they were often losers in the market economy, for the following reasons: (1) agricultural productivity increased rapidly and persistently; (2) because agriculture is highly competitive, the prices of agricultural goods decreased roughly in inverse proportion to rising farm productivity; (3) because the price elasticity of demand for agricultural goods is less than 1, the quantity of agricultural goods sold did not increase sufficiently to offset the decline in price, leaving decreases in farm revenues; (4) because the income elasticity of demand for farm products is less than 1, even if their prices were stable, farm incomes would lag behind the incomes of people in general; and increasing income in the population as a whole will not offset the deterioration of their revenues as a result of the low elasticity of demand.* On the other hand, farmers seem to have bargaining power proportionate to their numbers, if not greater, so farmers have been among the receivers of transfers in most industrialized countries in the twentieth century. These transfers have often taken the form of subsidies to agricultural production, which all economists agree are quite inefficient, or they have taken some other inefficient forms – and the taxes used to pay for them are also inefficient.

In general, then, the more the distribution of income is shifted away from the market equilibrium, the greater the cost from inefficient policies is likely to be. The inefficient transfers are chosen because their informational cost is less than the informational cost of making efficient lump sum transfers – and indeed the informational cost of such lump sum transfers might be infinite. This increase in inefficiency that goes along with a shift in the income distribution has been called "the big trade-off" for market economies by Dr. Arthur Okun.(4) It was a central idea of mid-twentieth-century welfare economics.

Deviations from Efficiency: Public Goods

Adam Smith wrote that:

> a duty of the sovereign or commonwealth is that of erecting and maintaining those public institutions and those public works, which, though they may be in the highest degree advantageous to a great society, are, however, of such a nature that the profit could never repay the expense to any individual or small number of individuals, and which it therefore cannot be expected that any individual or small number of individuals should erect or maintain.(5)

* The price elasticity of demand is the absolute value of (percentage change in quantity sold)/(corresponding percentage change in price), assuming other influences on the quantity sold are given. Thus, if the elasticity is less than 1, every 1 percent drop in price corresponds to less than a 1 percent increase in quantity sold, and accordingly a decrease in the total revenue of farmers. One other influence on the quantity sold is the income of the customers. The income elasticity of demand is (percentage change in quantity sold)/(corresponding percentage change in income). Thus, if that elasticity is less than 1, then a 1 percent increase in the incomes of buyers of farm goods would increase the sales of farm goods by less than 1 percent. In the first half of the twentieth century, economists made many studies of these elasticities and concluded that these elasticities are less than 1. Thus, the incomes of farmers would decline as the productivity of agriculture increases, a tendency that has indeed been observed in the twentieth century.

In the neoclassical economics of the twentieth century this idea was captured in the concept of a public good. Here again the ideas of Paul Samuelson were central.[(6)] We have already, in Chapter 2, given some consideration to public goods and to the idea that knowledge can be a public good. To reiterate, then, a public good is one that is non-rival and non-exclusive. A good is non-rival if one person's use of it does not deprive another person of the use of it. A good is non-exclusive if there is no practical means by which non-payers can be prevented from getting equal benefit from the good. A twenty-first century example of a pure public good, in this sense, is global satellite positioning services.

In a world of free information all the beneficiaries of a public good would get together and form a coalition to supply the public good, and agree among themselves on how to share the cost. This may in fact sometimes be done by a small group, through a cooperative arrangement among themselves. However, in the grand coalition of the whole society, the informational cost of doing this would be enormous. As with redistribution, as we discussed in the previous subsection, people making non-cooperative decisions would have incentives to hide the truth, misrepresenting the benefit they could get from the public good in order to reduce the share of the cost they might be asked to bear. The consequence most important for this chapter is that the public good would not be provided by non-cooperative decisions under the rules of the market game. Thus, if it is to be provided at all, it would have to be provided by some agency of the grand coalition as a whole, that is, a government.

However, there seem to be few instances of a *pure* public good as we have defined it. As a minimum, Smith mentions "public institutions and public works necessary for the defence of the society, and for the administration of justice." But, in particular, courts of law for the administration of justice are neither non-rival nor non-exclusive. It is quite possible to charge for the services of courts of law, and use of the law courts by any one individual contributes to their congestion, thereby (to some extent) depriving others of their use. However, the rules of the game for a market economy require that property rights and contracts are universally and impartially enforced, and this implies universal access to courts of law on an equal basis. Even if courts of law are not public goods, equal and universal access to courts of law does define a public good. Exclusion of non-payers would mean that access is not equal and universal, and maintenance of the condition of equal and universal access would require that the services be provided at a level that would limit congestion, so that the access of one person to courts of law, however congested, does not deprive another of it.

What we see here is that a judgment that access should be equal and universal transforms what would not otherwise be a public good into a public good. The same could be said of postal delivery and medical care. If the grand coalition of the whole society should determine that some standard of medical care should be equally and universally accessible, then that standard of medical care becomes a public good.

There may be many goods that *approximate* public goods, although they do not *strictly* fit the definition of public goods. These are sometimes called quasi-public goods. The case of streets and highways is instructive. They are neither non-exclusive nor non-rival. For example, toll gates can be erected to collect tolls and exclude those who do not pay, so that highways are not non-exclusive. Once they become congested, neither are they non-rival, since an additional user of the highway increases the congestion and thus deprives other drivers of the opportunity to use it *at the state of congestion* that might exist otherwise.

In any case, public and quasi-public goods provide an important class of exceptions to the norm that non-cooperative decisions within the rules of the market game would be efficient. Non-cooperative decisions on production of these goods would not be efficient.

Deviations from Efficiency: Externalities

One of the key ideas originated by Pigou's founding book on welfare economics was externality. We have already discussed externality in the context of the cost of information in Chapter 6. All that needs to be added at this point is that decisions about activities that generate strong externalities will not be made efficiently by non-cooperative decision-makers in the context of the rules of the market game. Small groups may be able to solve these problems by cooperative agreements among themselves, and local groups acting within customary norms often have done so.(7) Nevertheless, externalities provide many instances in which non-cooperative decision-makers acting according to the rules of the market game will not make efficient decisions. If externalities are to be controlled efficiently, this will require some action by an agency of the grand coalition of the whole society itself, that is, the government. One possibility that Pigou pointed out is that the activities that produce externalities might be efficiently priced, with the efficient prices enforced through a system of taxes and subsidies.

Competition Policy

In the world as neoclassical economics envisions it, then, the market game would satisfy criterion 1 in the listing in Chapter 7. However, it might not satisfy criterion 2. A conspiracy in restraint of competition is a cooperative deviation at the expense of others in the grand coalition of society. (Namely, it is at the expense of the customers of the companies that enter into the conspiracy.) Thus, the market game may require some policies against the formation of such deviations, that is, against collusion and conspiracies in restraint of trade.

Even in the absence of collusion, monopolies might be established for a number of reasons. In the neoclassical welfare economics of the mid-twentieth century, it was believed that this in itself would lead to inefficiency, since the monopoly would have both the power and the incentive to raise its price above the equilibrium price. This would be inefficient in the same way that a tax on the product of the monopoly would be inefficient: it would result in a reduction of the quantity bought below the efficient quantity. Thus, welfare economics argued that government policies might be needed to prevent monopoly, or to correct the inefficiencies that might arise from monopoly. However, in the light of some of the discussions in the previous chapters of this book, this monopoly theory and policy seems to have been misconceived.

The neoclassical monopoly theory is based on the law of one price, that is, the assumption that the monopoly will not engage in any price discrimination. But why would it not? In some cases, there may be a secondary market for the item sold by the monopoly, which would frustrate any attempt by the monopoly to practice price discrimination. Those who get a discriminatory low price can then resell their purchases to those who would otherwise face discriminatory high prices, with the result that the monopoly would sell nothing at the higher price. An example is the sale of tickets to performing and sports events. In this case resale is called "scalping" and is sometimes illegal. The performance monopolies commonly support these anti-scalping laws. But in the absence of any secondary market, a monopoly would almost certainly find it profitable to practice some price discrimination, and would do so.

Moreover, the neoclassical theory of monopoly ignores the fact that exchange is a cooperative relationship. The monopoly seller and the buyers are members of a cooperative coalition, and it is in the mutual interest of all of them to produce and sell the efficient output.[(8)] Further, while the establishment of a monopoly will shift bargaining power away from the customers, it seems likely that they will still retain some bargaining power, and so retain some of the surplus created by the coalition, even if they do not retain the consumer's surplus that they would obtain in the absence of monopoly.

Nevertheless, let us suppose that the buyers have no bargaining power at all, and further suppose that information is free. Then the monopoly would practice perfect first degree price discrimination, selling (for example) every ticket at the highest price that anyone anywhere would be willing to pay. But in the actual world the information cost of first degree price discrimination would be very great. In a world of costly information, then, the monopolist would practice cheaper second degree or third degree, price discrimination,* or a combination of the two, which would leave the buyers with some surplus from the trade. Putting that point otherwise, we see that information cost conveys some bargaining power to the buyers. It follows that second and third degree price discrimination are to be expected as a rule, apart from special cases.

Moreover, in a world of costly information, the distinction between monopoly and non-monopoly is blurred. Since there is an information cost of recruiting new customers, each seller will act somewhat as a monopolist toward those enrolled in that seller's coalition. The seller will commonly practice price discrimination, calling it a "customer loyalty program" or some such thing.

In any case, there is little reason to think that monopoly power would result in restriction of sales below the efficient output, except in special cases such as scalping. On the other hand, monopoly power will shift bargaining power away from the customers of the monopoly, and thus will affect the distribution of benefits from cooperation in the grand coalition of the country as a whole, so public policies that would prevent monopoly or limit monopoly price discrimination might be less costly in informational terms or less inefficient than measures to redistribute the monopoly profits via taxes and transfers.

Interim Summary

In this section, with one qualification, we have reproduced the neoclassical welfare economics that was developed with great precision and logical power in the mid-twentieth century. These ideas can be subsumed readily in the framework suggested in this book: that economics is about cooperative action, imperfectly realized in a world of costly information. The one qualification is that neoclassical economics seems to have misunderstood monopoly, lacking, as it did, any clear conception of cooperative action other than bilateral exchange.

* In third degree price discrimination different prices are charged only between different markets, which presumably can be identified as such at a low informational cost. In second degree price discrimination, by contrast, an individual buyer may pay different prices for different units. An example of this would be a discount for volume purchases.

SOME QUALIFICATIONS

In this section we consider three issues for the market economy game that arise in a world of costly information but that were not part of the consensus of mid-twentieth-century neoclassical economics.

Coordination, Again

In the first section of Chapter 7, the focus was on coordination within the coalition – whether determined by hierarchy, voting, custom, or otherwise. However, we now envision a world of many coalitions, and problems of coordination may also arise among the various coalitions. Certainly there are institutions intended to resolve these problems, and they are quite familiar: contracts, exchange, and markets. Neoclassical economics describes a world in which markets coordinate the decisions of the various coalitions (business firms) in a way that, if not efficient, deviates from efficiency only in predictable ways. But neoclassical economics describes a world in which information is free and human beings are "absolutely rational decision-makers whose capabilities of reasoning and memorizing are unlimited."(9) Since information is costly and real human rationality is bounded, decisions are made according to heuristic rules that only approximate the efficient cooperative solution for each coalition. We might hope that a set of heuristic rules that approximates the rational decisions for each business firm or other coalition for production and exchange would also approximate the coordination that would exist among the various coalitions that make up the market economy that we would expect in the ideal world of free information. Unfortunately this is not true.(10)

Suppose, for example, that a decision-maker believes that the costs of his or her firm are well approximated by the sum of a constant overhead cost, F plus prime cost, that is, the cost of labor and materials; and that prime cost is roughly a constant, v, times output. Then cost is approximated by $F + vq$, with q representing output. The decision-maker then sets his or her price at a constant mark-up over v and sells "whatever the market will bear" at this price. If F, v and the mark-up are well chosen on the basis of experience, the result will be a good approximation to the profit-maximizing price and output under routine conditions. However, an economy inhabited by firms that operate according to this rule of thumb will function roughly in the way described by Keynesian economics, and not at all in the way described by neoclassical economics.

What does this have to do with coordination? For the most part, this is a topic for macroeconomics, which we reserve for the next chapter. However, here is an example more usually applied to the economics of less developed countries. Consider the investment game at Table 8.1. For this game the investors are A and B, and each can choose among three strategies: don't expand, expand a little, or expand a lot. If one chooses to expand, his or her investment creates a demand for the product the other one produces, but if one expands and the other does not, the one who expands will not find a market for his or her expanded production. Payoffs are relative profitability on a scale of 1–5 (with –5 for losses). We see that this game has Nash equilibria whenever both investors choose the same scale of investment. It is a pure coordination game. This implies (1) that any of the three levels of investment may be stable, if both choose it, (2) that all three strategies are rationalizable, so that (3) if an investor relies on rationalization to make his or her decision, he or she cannot rationally decide among the strategies without some further information. As we saw in Chapter 3, rationalization may lead to the less profitable equilibrium, or, indeed, to a non-equilibrium outcome at which just one invests. In the context of this chapter, we recognize that the cooperative decision is for both to expand a lot. However, to come to this decision, each individual decision-maker must produce information to answer two key questions: "How will my profits depend on the other firm's decision?" and "What will the other firm decide?" In a two-person game this information could be pretty cheap – they could just talk to one another. But what our hypothetical grand coalition of the whole society faces is an investment game among millions of market decision-makers. The cost of this necessary information, to any one decision-maker, is likely to be more than the potential profits from the investment in the best of cases. It would not be worthwhile for the investor to attempt to make an informed decision. Instead, the investors could be expected either to minimize their risks by not investing, or, if they invest, to do so on the basis of "animal spirits" or "irrational exuberance."

Table 8.1 An Investment Game

First Payoff to A, Second Payoff to B		B		
		Don't expand	Expand a little	Expand a lot
A	Don't expand	1,1	2,0	3,–1
	Expand a little	0,2	3,3	4,2
	Expand a lot	–1,3	2,4	5,5

It seems that a world of costly information is a Keynesian world, or perhaps something even stranger and more complex than that. Those who advocated for neoclassical economics against Keynesian economics in the mid-century period often described informational costs and such as frictions, and argued that the frictionless neoclassical model would approximate reality as frictionless models in mechanics supposedly do. But the previous two chapters of this book have argued systematically that informational costs are not frictions but fundamental explanatory principles that account for the existence of such things as money, organizations, and externalities. In a world of costly information, problems of coordination such as those central to Keynesian economics must be taken seriously.

Asymmetrical Information

As we have seen in Chapter 5, markets for medical insurance can be inefficient due to asymmetrical information. Asymmetrical information exists when different decision-makers have different information, and some decision-makers have information that other decision-makers need to make efficient decisions. In a world of cooperative decision-making and free information, this would not occur – the needed information would be produced at no cost and used to make decisions that are efficient from the point of view of the whole coalition. However, when we rely on noncooperative decisions – the rules of the market game – inefficient decisions may result. Among the consequences are moral hazard and self-selection at the equilibrium (both discussed in Chapter 5), both of which can be inefficient. It should be stressed that there may be no perfect solution. The cost of producing the necessary information for cooperative decisions may be great. An incentive-compatible set of rules, which would lead people to reveal the information they have in their own interest, may conflict with other objectives of the market game, and may be costly to implement. Thus, it may be that there is no real possibility to improve on the inefficient equilibria that arise due to moral hazard and self-selection.

But defeatism is a mistake. Asymmetrical information was a discovery of economics in about the last quarter of the twentieth century, and was honored by the Nobel Memorial Prize of 2001.(11) Along with this discovery came a number of proposals by which government policy might improve on the market equilibria in the context of asymmetrical information. In the context of this chapter, what we have learned is that a grand coalition of the whole society might choose to deviate from the rules of the market game, directing its agency, the government, to impose specific regulations, taxes, and transfers, to improve the efficiency of the market allocation of resources in the presence of asymmetrical information.

Entrepreneurship

Thus far, we envision a market economy with a given set of coalitions for production and exchange. This might be described as a short-run view. However, business firms do not exist forever, and still less do they exist from time immemorial. The coalitions for production and exchange in our market economy are rather a population, continually depleted by death and renewed by birth, and this needs to be incorporated in the vision of the grand coalition of society as a whole. For this purpose, we need to return to entrepreneurship, in some of the many senses of that word.

In Chapters 3 and 7, we have followed John Bates Clark's understanding of the entrepreneur as the coordinator of complex combination of labor in production. Chapter 6, by contrast, stressed the Austrian and Schumpeterian concepts of entrepreneurship that focus on the creative entrepreneur,(12) who founds a new business. It is clear that the initiative to establish a new coalition – be it a new business, workers' cooperative, non-profit corporation, church, club or family – can only be taken non-cooperatively. That is, until the coalition is formed, there is no group in the interest of which cooperative decisions might be made. However, the act of creative entrepreneurship changes the world, and from that point forward, the activity of the entrepreneur takes place in a cooperative framework. Perhaps this is the most difficult transition for many entrepreneurs!

Creativity is little understood in economics.(13) Among those who study creativity in psychology and related fields, the consensus is that the creative product, viewed in retrospect, is recognized both as adequate to some known purpose and as surprising, as something that would not have been anticipated before it is observed. In the language of Chapter 6, creative information production provides information on the unknown unknowns. It is surprising in that it brings the unknown unknown into our knowledge, and is adequate in that it resolves the unknown unknown.

Can a creative act take place within a cooperative framework, or is creativity intrinsically individual? Here is a case that it is intrinsically individual: cooperative arrangements are directed toward mutual benefits *that are foreseen by the cooperators*; and since creative acts generate conditions that were not foreseen, they cannot be the objectives of the cooperative group. But this argument needs careful interpretation. It would be valid as given in a world of free information, in that in such a world, the cooperative agreement would specify every detail of each agent's actions, and indeed this would leave no room for the individual to act creatively. But, of course, in that world there would be no unknown unknowns – no scope for creativity in any case. In real organizations, the individual may choose his or her own course of action, cooperatively or non-cooperatively as the case may be, within the rules of the game. Thus, an individual may often have scope for creative activity that will enable the coalition to better provide mutual benefits to its members. But, conversely, this provides another reason for cooperative groups to tolerate non-cooperative activity within broad rules: such tolerance may give rise to more creativity that leads to greater benefits all around.

In particular, this is a sketch of the role of entrepreneurship in a market economy. By tolerating and indeed encouraging non-cooperative acts of entrepreneurship, the market economy encourages the growth of new coalitions of production and exchange and increases the overall production in the economy. This, too, however, will influence the distribution of benefits in the economy. Presumably the entrepreneurs will get the bulk of them, and this may not advance the cooperative objectives of the grand coalition of the whole society, reinforcing the importance of "the big trade-off" between efficiency (and entrepreneurship) and equality.

Discussions of entrepreneurship often presuppose that the entrepreneur is motivated by profit-seeking. This is often true, but creative freedom can be rewarding in itself, and the satisfaction of accomplishing something – Veblen's "instinct of workmanship"[(14)] – also plays a role, and these may complement and reinforce the motive of profit-seeking. Moreover, creativity does not take place in a vacuum of information, and in many cases, markets provide information both on what might be successful, before the creative act takes place, and evidence of its adequacy after the creative act has taken place. Thus, non-cooperative action within the rules of the market economy provides a favorable environment for entrepreneurship. But market information may be misleading, as we have seen above. Where that is so, profit-seeking will not lead to adequate solutions. But entrepreneurship can arise from other sources. It may arise from philanthropic motives in the case of social entrepreneurship, and instances of government entrepreneurship are not hard to find. What is crucial is that opportunities for entrepreneurial creativity are very widespread in society and the creative action is allowed to work itself through, and this will require a great deal of tolerance for non-cooperative decisions.

Interim Summary

We have supposed that a hypothetical grand coalition of the whole society might establish a set of rules within which non-cooperative decisions would be relied upon for most or all day-to-day economic decisions. This has two important advantages. First, to the extent that individual willingness to pay reliably reflects the benefits the individual derives from the economy, a system of property rights and enforceable contracts will encourage producers to direct their production in ways that best benefit the group, as Adam Smith famously pointed out. Second, it provides a friendly environment for profit-seeking, creative entrepreneurship, which leads to the formation of useful new cooperative arrangements among the individuals who form the society. However, there are several ways in which this arrangement may not achieve the best cooperative outcome. First, the market game determines the distribution of benefits among the members of the grand coalition, and this may deviate from the distribution the grand coalition would choose. This leads to the establishment of an agency of the grand coalition, which we may call a government. The government then may redistribute benefits by means of taxes and transfers, and because information is costly, may choose to do this by policies that are inefficient but less costly in informational terms than formally efficient policies would be. Second, government must take the responsibility for the supply of some public goods, including the enforcement of the rules of the market game itself, and may be able to improve the efficiency of the market game by deviating from it in the cases of externalities and asymmetric information. Finally, since information is costly, coordination of decisions among the various decision-makers in the market game may fail, resulting in recessions and depressions, or even long-term stagnation, and interventions by the government may improve coordination.

VOTING, AGAIN

For this chapter, we have explored the implications of some ideas from the previous chapters for a hypothetical grand cooperative coalition of the whole society. One possibility is that the grand coalition might rely on the rules of a market economy, and thus on non-cooperative decision-making, at a much lower informational cost than a more centralized system. But we have observed that a market system is only incentive compatible in very special circumstances that cannot be realized. That is, in terms of twentieth-century neoclassical economics, the market economy fails to produce an efficient allocation of resources if there are public goods and in some cases of externalities, monopoly, and asymmetric information. At the very least, uniform protection of property rights and contract enforcement must be provided, and constitute in effect a public good. Moreover, the distributional outcomes of a market system may not be satisfactory from a cooperative viewpoint. All of this was well known to twentieth-century neoclassical economics. The implication is that, in our hypothetical grand coalition, there will be a role for a hierarchy. Moreover, the details of the laws, the particular system of property rights, and the details of the tax system must be decided. Here is one possibility: let the market system in economics be joined to a system of elections and voting to determine the hierarchy and to select those people who will determine the rules of the game; and then rely on non-cooperative buying, selling, and voting to obtain an approximately cooperative outcome. However, there are some real difficulties with this vision.

Voting Mechanisms

Because of its importance in modern political systems, voting has been studied by mathematicians and social scientists since the 1700s, and economists and game theorists, in particular, have contributed important ideas. Voting is not just one procedure – there are a number of voting rules that have been put into practice at different times and places and still others that have been proposed but never yet tried. For example, voting may be based either on plurality rule or on majority rule. Plurality rule means that the candidate or alternative that gets the largest vote is the one chosen, while majority rule means that a candidate or alternative is chosen only if she or he gets more than half of the votes. When there are three or more alternatives, it may be that none is supported by more than half of those who vote, so majority rule may not give a decision – and that contradicts criterion c from the Baumol and Quandt list of characteristics of rules for optimally imperfect decisions (Chapter 6 of this volume). Thus, majority rule needs to be extended in some way to be sure we get a decision. In *Robert's Rules of Order*, the widely used American manual of committee procedure, the requirement is that the voting be repeated until a majority is obtained. In many political jurisdictions, where this sort of persistence is not workable, the rule calls for a single run-off election between the two candidates who have obtained the most votes in the first round. In other elections – such as British Parliamentary elections – the candidate or alternative that gets the larger vote simply wins, whether the vote is a majority or not. This is plurality rule. Majority rule has been more widely studied, and is simpler in some ways, so this section will refer to majority rule with one explicit exception.

There are, of course, information costs of voting. Voters have to inform themselves as to what the alternatives are, and determine which alternative to vote for. The votes have to be tallied and counted and this, too, is an informational cost. At the same time, voting is itself a process of information production, since each person provides the information as to the alternative the voter favors; and this may be a cheaper method of information production than alternatives, such as multilateral bargaining among the whole electorate. We might describe voting as an information filter that takes the information available to the voters and transforms it into information about the preferences of the voters.

Non-cooperative Voting

The theme of the previous chapter and this one is that a cooperative coalition may choose a set of rules and then rely on non-cooperative decision-making within those rules to attain its cooperative objectives. Suppose the group adopts a rule that all group decisions will be made by voting, with the vote determined non-cooperatively. Will this attain the objectives of the group? Even if the agents are perfectly rational, and information is free, the answer is no. In general, voting is not incentive compatible.

If people vote non-cooperatively, they will often vote strategically. That is, a voter may often reason that if she votes for her first preference, she will actually get a worse result than she would get if she votes for a second or third preference. For example, a voter who would prefer an environmentalist candidate might instead vote for a candidate who is moderate on environmental issues, in the expectation that voting for her first preference would divide strong from moderate environmentalists and so result in the election of an even worse candidate for the environmental issues. From twentieth-century research in game theory, we know that there is no voting procedure that will lead non-cooperative voters always to choose their first preference.(15) If, then, we think of voting as an information filter, this means that the output of the filter is somewhat unpredictable. The output may be the voter's first preference, or it may not. The unpredictability of the output means that the information will be less useful than it might otherwise be, and may not be sufficient information on the basis of which to make efficient decisions for the coalition as a whole.

There are two further difficulties with a non-cooperative interpretation of voting. One, which does not seem to have been studied, is that strategic voting may result in multiple Nash equilibria. Suppose there are two or more candidates or issues that are close substitutes in the preferences of a majority faction. There is a danger that the majority might divide its votes among the substitutes, so that none gains a majority. This will not be a Nash equilibrium. Conversely, there will often be Nash equilibria where the majority faction all vote for any one of the substitutes.(16) If two or more such Nash equilibria are Pareto-preferable to non-equilibrium outcomes, then this is a coordination game, and in the absence of any further information or communication, rationalization may lead the voters to a worse result than any Nash equilibrium. Coordination problems of this sort must be common when there are two or more candidates or legislative measures that are close substitutes in the minds of many voters. In such a case, to obtain a Nash equilibrium, the voters will need some signals to suggest to them one Nash equilibrium rather than another. Endorsements by a political party or a prominent leader may provide the signal. This is one function of political parties, but this is, of course, only part of the story.

In any case, the non-cooperative interpretation of voting cannot be complete. Voting imposes some cost on the voter, if only a time cost; and in large-scale elections there is negligibly little chance that a single vote will decide the election. Thus, the private benefit from voting is negative; people who make non-cooperative decisions will not vote! In some countries, in recognition of this, voting is compulsory and those who do not vote are fined. But it does not seem to be feasible to require voters to inform themselves. In some other contexts we may bring the voters together in a meeting, in which voting and discussion to enrich the information content of the vote are relatively cheap; but this town meeting democracy will also not be feasible in many cases, and going to a meeting also has a cost. On the whole, the decision to vote can only be a cooperative decision. Conversely, the fact that many people do vote is perhaps the most pervasive evidence that real human beings do have some capacity to spontaneously make cooperative decisions, even in the absence of enforcement.

One further point should be made, here. For some people, choosing leaders by voting, particularly by majority rule, is a value in itself, at least in some governmental organizations. Thus, even if it were not efficient, if, for example, the informational costs of democratic processes are greater than the benefits, we might wish to choose leaders by voting nevertheless.

Cooperative Voting

Most studies of voting in game theory and economics have focused on whether voting is incentive-compatible and thus on non-cooperative equilibria of majority rule games. These equilibria are typically inefficient. On the other hand, if voting blocks are considered as cooperative coalitions, no winning coalition is stable. Consider, for example, a simple three-person game among Andrea, Bill, and Carol. Any coalition of two of them can divide a payoff of 1 between them in any proportion that they choose. Suppose, then, Andrea and Carol form a winning coalition and divide the payoff equally. Then Bill can approach Andrea and offer to form a new winning coalition in which he will settle for 1/3. This will make both Bill and Andrea better off. But now Carol can approach Bob and propose to coalesce with him and divide the payoff equally, making both Bill and Carol better off. In this way, every winning coalition in a majority game is dominated by another winning coalition. This is called a dominance cycle.

But, again, both of these views overlook two things. First, if people are rational, they will anticipate this dominance cycle and not go pursuing gains that will never be realized. (Even boundedly rational human beings are likely to realize this sooner or later and to adopt heuristics that will lead to a more stable outcome.) Second, in all of this we are supposing that voting is a way of settling differences within a coalition already formed. As with the auction, the organization of an election is itself a cooperative act, with a set of commonly agreed rules and procedures: this presupposes an existing coalition of all the voters. Why then would such a coalition choose majority voting as a mechanism for its joint decisions?

Voting as Bargaining

The procedures of many legislative bodies and committees seem better described by a two-stage game. At the first stage there is negotiation, and then at a second stage, a vote is taken. Bargaining power will play a role in the negotiation at the first stage. We assume (as argued in Chapter 3) that bargaining power arises from threats. If the disagreement outcome is that a contested election will be held, then there is a relatively limited range of threats that can be made. One can threaten to vote against an alternative that another participant very much wants, but one cannot (within the election game) threaten to beat him up or put a bomb in his mailbox. This will have an effect on the bargaining power within the coalition. As these threats are somewhat symmetrically distributed, bargaining power in the negotiations at the first stage should be roughly equal. Perhaps elections are often chosen as a decision mechanism for just that reason.

This would be a theory of representative democracy. We have stressed that a conference and negotiation are not feasible for a large population, because of the information cost of such a negotiation.(16) Voting in the general population is most commonly for representatives, not on issues, though referendums are possible and fairly common in a few American states. The candidates typically represent different factions or political parties, and voting is largely motivated by factional loyalty. When representative government is successful, however, the candidates anticipate that negotiation and compromise will be the business of government, and that the division of power will have to be renegotiated after yet another election. Direct democracy is a popular ideal, but too much direct democracy can be self-frustrating, as we have seen from time to time in the United States. The exception to this is town meeting democracy, where the whole population (of a small town) can be brought together to play the two-stage game of negotiation and voting. Gathering a group in one location radically reduces the cost of information and seems essential for negotiation.

While there has been little study of joint models of negotiation and voting, a little is known. First, if the negotiators are rational, and information is free, the outcome will be efficient, at least with respect to the preferences of those participating in the voting. Second, if the game is a transferable utility (TU) game, (See Ch. 3) the surplus generated by the coalition is divided equally. In a world of free information, lump sum transfers are possible, and that would seem to make the game of negotiation and voting a TU game. If the game is a non-transferable utility (NTU) game, the division is not generally equal, but nor does it correspond to any simple scheme such as compensation according to product. These conclusions are drawn from a model that is symmetrical in that any possible winning voting block may be formed.[(17)] When some winning voting blocks are excluded because of party loyalty or for other reasons, the bargaining power of the factions is affected unpredictably, and the distribution of gains may be unequal even in a TU game. For a variety of reasons, the preferences of representatives may not correspond precisely to those of the general population, and this could lead to outcomes that are not efficient from the point of view of the general population. Further, in a real world application, costly information and bounded rationality are likely to result in further deviations both from efficiency and equality. Negotiations may fail, and voting threats may be carried out.

Nevertheless, we may hope that the results of the preliminary negotiation will be an efficient compromise among the factions represented in the legislative body. Suppose, for example, that the Farmers' Party has the majority and it pursues this advantage by proposing policies that are inefficient, even after taking informational costs into account. To say that they are inefficient in this sense is to say that the constituents of the other parties are harmed, and the costs they bear are greater than the benefits gained by the farmers. It would be reasonable for the other parties to offer some side payment to the Farmers' Party to obtain some adjustment to reduce their own costs. If there is no such offer that the opposition parties can make that is feasible in terms of its own informational costs, then we must conclude that the original proposal was efficient after all, *allowing for information costs.* If there is such a feasible offer, then presumably it will be made and accepted. Further, even when parties in a legislative body act non-cooperatively, they interact again and again. This repeated game creates the potentiality of a cooperative solution among the parties. This will be reinforced by the expectation that other parties may have the majority after a future election. Conversely, permanent majorities and one-off general elections are not associated with successful democratic government.

Voting in the Population at Large

To say that the individual's decision to vote is a cooperative decision is to say that it is a decision aimed at gaining a mutual advantage for the members of some group. What group? In fact, of course, coalitions such as parties, groupings of parties, and political clubs are common enough in electoral politics that the word "coalition" is derived from electoral politics.

Suppose I try to make my voting decision on an ethical basis, in a Kantian vein, as best I understand it. Then I will attempt to frame a rule for my voting decision such that I might wish everyone to act on that rule. As a beginning, consider the decision whether or not to vote. This does not seem very hard. The rule that suggests itself is that everyone should vote, despite the personal costs and inconvenience, unless extraordinary circumstances were to make it impossible or much more costly. And what should I vote for? If there were one alternative that is clearly best for the whole group, then no doubt I should vote for that and will everyone else to do so, but in general there will not be. This will be clearer in a specific example. Let us suppose that I am a member of a food cooperative. I might prefer that the store use more of its shelf space for prepared food, since I am often in a hurry and don't want to cook, while some other members might prefer that the space be allocated to fresh fruit. Neither decision is objectively best for our whole group. An ethical rule for voting would have to recognize that different people may simply have different preferences about some decisions of the cooperative coalition. Instead of attempting to frame a rule for my voting decision such that I might wish everyone to act on that rule, I might instead form a rule according to which every voter *who shares my preferences* would vote. In the case of the food cooperative, I could affirm that members like me, who are commuters with little time for cooking, should vote for products that serve our needs, and so personally feel comfortable voting for more space for prepared food. The rule, that is, seems to be that each should vote according to the interests of her or his faction.

Now, this is a bit puzzling – faction does not seem an appropriate category for an ethical rule, on its face. Nor is a faction a clearly defined term. In the example of the food cooperative, there may be some harried commuters who are not willing to give up fresh fruit to have more prepared food, but want more of both and the space reallocated instead from shelves now devoted to boxed breakfast cereal and soft drinks. Are they of my faction, or not? This could be a strategic decision – are we prepared-food lovers numerous enough to win without the fruitarians? Or would we be better off sharing a factional identity with the mothers of clamorous children who want more choices of breakfast cereals and fruit drinks?

But we are not really concerned with ethical rules, here. Rather, we are concerned with heuristic rules that might be used by boundedly rational voters to decide their votes among a limited slate of candidates to be their representatives or among a limited set of alternatives available to a cooperative group of which they are part. Cooperation is about the generation of mutual benefits, but the benefits are distributed according to bargaining power. By supporting my faction, I enhance its bargaining power and thus my own. Within a cooperative group, a faction is a cooperative deviation in the sense of condition 2 in the second section in the previous chapter, an attempt by a subgroup to cooperate among themselves (in voting) to the disadvantage of other members of the group. But this needs not conflict with the cooperation of the group as a whole. To the extent that the voters send representatives to a legislative body who will then play the two-stage negotiating and voting game discussed in the previous subsection, we may hope for some efficient settlement among the factions.

The point is that horse-trading among factions, like horse-trading among frontiersmen, tends to eliminate inefficiencies. Thus, voting is not a case of a cooperative group relying on non-cooperative decisions, within the rules, to realize its cooperative objective, as proposed in the first section above. Rather, it is a case of a cooperative group permitting cooperative deviations, within the rules, to generate decisions that are approximately cooperative. This relies on individuals to make cooperative decisions (to inform themselves and vote) that are privately costly to them, perhaps because of factional loyalty. Obviously this arrangement, like most human arrangements, is fallible and can fail in different ways. Perhaps we could think of a voting process as one that enlists cheap cooperative behavior – voting – which at least a large proportion of the population will undertake without rewards or penalties, to bring about enforcement of other cooperative strategies, such as making an efficient effort in production or paying taxes, that people often will not undertake without rewards and penalties.

Interim Summary

In summary, then, it appears that a grand coalition of the whole society that would rely *primarily* on market rules in economic matters and on representative democracy in political matters, tolerating non-cooperative decisions in markets and cooperative deviations by entrepreneurs and political factions, will provide a better approximation to cooperative decisions than other systems that have been widely tried. To that extent we might consider the grand coalition of the whole society as a possibility, if by no means a certainty.

ECONOMIC PLANNING RECONSIDERED

After the experience of the twentieth century, and after all that has been said in this chapter, why would one want to reconsider economic planning? The point is that the advantages attributed to a market system in this chapter are abstract: market allocation is approximately incentive-compatible in some circumstances and is open to organizational creativity. At best, market organization is a *sufficient* condition to realize these advantages, not a *necessary* condition. Further, the conditions under which it is a sufficient condition – absence of externalities, public goods, and coordination problems – are remote enough from reality that one may reasonably wish to consider some alternative set of sufficient conditions. This section sketches a scheme of economic planning that would approximate incentive compatibility (whenever a market system would) and remains open, in principle, to free initiative. Further, it argues that the planning mechanism proposed could be more robust to externalities and to coordination problems than a market system can be, and would allow for a more flexible distribution of the net benefits. While it draws some ideas from Sir Arthur Lewis's market socialist schema,[(18)] this will be a "socialist" schema only in the limited sense that it requires predominant government ownership of the non-human means of production, and many details would be roundly criticized by many socialists.

In planning, it is important to distinguish between the formation and implementation of a plan. For command-control planning, implementation is trivial – "a plan is a command" – but this very fact makes it impossible to *form* an efficient plan, since the necessary information will systematically be concealed. This fact is well understood in economics and probably is a major factor in accounting for the poor results of planning in the Soviet Union in the twentieth century. Command-control planning also seems to offer little or no scope for Schumpeter's "creative response," and thus to create a barrier to useful innovation.

By contrast, indicative planning in capitalist economies has no means of implementation other than announcement effects. McCain[(19)] has argued that announcement effects can make a difference, but in practice indicative planning has not been even as successful as command-control planning. In a market socialist society along the lines proposed by Lange[(20)] in which the plan bureau would play the role of an auctioneer in a Walrasian market economy, implementation would be trivial, since producing the planned quantities for sale at the planned prices would be in the interests of the decision-makers in the enterprises. However, we are concerned here to find a system that might improve on the market equilibrium, in the light of its shortcomings; so that market socialism is no solution. Nevertheless, market socialism suggests the solution to the problem of plan formation and implementation: the plan should be formed in such a way that the decision-makers in the enterprises, responding to the pressures and exigencies of their own situations (and using the knowledge and creativity that they possess and that may not be available to others, including the central authorities), will choose to implement it.

The Plan: Simplest Case

Following the outline of the market socialist proposals we envision an economy in which production of goods and services is carried on by autonomous local organizations that sell their output at prices that cover the private opportunity cost of production. These organizations are corporations in the sense that they are capitalized through shares and their gross profit is returned to their shareholders as dividends. However, most if not all of the shares are owned by a sovereign wealth fund. Further, for now we assume that savings comprise retained profits. Thus, the sovereign wealth fund commands most, if not all, of the savings available for investment in any particular period, and the economic plan is understood as a plan for the disposition of that fund.

We suppose that the plan will be based on some definite priorities given by the grand coalition of the whole population, via its instrumentality, the government. The simplest possibility is that these priorities are expressed in a social welfare function that evaluates the possible outcomes in terms of their efficiency and the equality of distribution of the net benefits of production. The key necessity is that the objectives of the plan are articulated in such a way that proposals for investment of a part of the available fund can be evaluated in terms of their impact on the objectives, as, for example, by distribution-weighted cost–benefit analysis.[(21)]

The second stage of the planning process envisioned here is the request for proposals. The proposals would be for expenditure of investment funds and might come from existing corporations or from agents that might wish to start up new corporations. Each proposal would include a provision for the replacement of the opportunity cost of the resources used, whether through the issue (to the Sovereign Wealth Fund) of new dividend-paying shares or as interest and amortization or some other arrangement.

The proposals would then be evaluated first as business plans, with respect to the reliability of the facts they assume and the outcomes they predict. Those that are considered satisfactory in that evaluation would be further evaluated on the basis of their incremental contributions to social value as defined in the objectives of the plan. Funds from the Sovereign Wealth Fund would then be allocated to the proposals with higher evaluations on this score, within the limits of the availability of funds. These proposals would become the economic plan for the period. Each proposal accepted and funded would become a contract between the Sovereign Wealth Fund and the enterprise that would ultimately be funded. Thus, implementation would simply be enforcement of the contracts, and since the decentralized decision agencies would not knowingly propose contracts that would leave them worse off on the whole, this should be sufficient. More importantly, they are able to draw on the information they possess, or might create, that would not otherwise be available to the planning bureau, in forming their proposals.

These procedures have some precedents, of course, and for this section that evidence of possibility is far more important than any claim of originality:

- The request for proposals and review of proposals as candidates for funding is familiar in the United States from research funding processes by agencies such as the National Science Foundation and the National Institutes of Health. These procedures are designed specifically to support the production of public or quasi-public goods in which creativity and some element of tacit knowledge are particularly important. They certainly are not perfect processes – it is sometimes suggested that the reliance on peer reviews, intended to assure the high scientific quality of the proposals, could inhibit major innovation. In any case, the proposal-first process is intended to be open to creativity, in an environment in which outcomes are inherently uncertain.
- The evaluation of business plans as a condition for funding enterprises in commodity production is a routine part of the resource-allocative activities of the United States Small Business Administration. In this case, of course, the funds allocated come from private sector lenders, although the loans are guaranteed by government; replacement takes the form of interest payments to the lender.
- The call for "shovel-ready" projects in the context of the 2009 fiscal stimulus response to the great recession is also suggestive of the planning process outlined here.

This is, of course, no more than a sketch, and there will be no shortage of criticisms and counter-arguments. We next consider some complications, and then reconsider the assumption that all productive organizations are corporations producing commodities for sale.

Some Complications

The previous subsection assumed conditions in which a market system would be incentive-compatible: absence of externalities, public goods, and coordination problems. We now reconsider those assumptions in part.

In many ways, externalities are simplest, since (to the extent that the facts are known) we know how to correct for externalities in a market system: Pigovian taxes.[22] The proposal-first planning procedure puts this in a somewhat different light. External costs are among the opportunity costs of a project, and so the norm that opportunity costs should be repaid will apply to external costs in particular. Accordingly, the contract between the proposing enterprise and the Sovereign Wealth Fund might specify (in addition to dividends and amortization of interest) a payment per unit of externality-generating activity. That is, the Pigovian tax might be included in the proposal and the contract and thus in the plan, rather than being imposed by legislation.

Coordination failures, that is, recessions and depressions, present a different sort of problem. Sir Arthur Lewis's thought is firmly grounded in classical political economy, and that is reflected in the understanding of investment in the planning proposal. If indeed profits were routinely saved and reinvested and there were no other saving or investment in the economy, "general gluts" presumably could not occur. This classical viewpoint is especially evident in Lewis's transitional program, which was that the government run a surplus of tax receipts over expenditures for an extended period, first paying off the public debt and then purchasing stock in corporations. Lewis's classical assumption was that the resulting increase in social saving would simply accelerate the growth of the national economy. If, as in a Keynesian world, people may predominantly attempt to increase their portfolios or their hoards of money, or both, producing a general glut (or in more modern terms, a recession) for a certain period, then we may not be so confident of the classical result. It will not be possible, in this brief discussion, to integrate the planning process proposed here with a model of macroeconomic policy, but it seems likely that an economy under this sort of planning would provide more degrees of freedom to the central authority to deal with economic fluctuations.

The Organizational Bestiary

Writing in 1950, Lewis took it for granted that corporations were not managed in the interest of shareholders but by managerial discretion, so he saw no need for nor benefit from the substitution of ministries of production for corporate management. In the twenty-first century, we may be less comfortable in leaving decisions to managerial discretion. Certainly, for the purposes of this proposal, corporations will have to be transformed in at least one way, namely that they would not be allowed routinely to retain profits. Instead the profits must be distributed as dividends. (Retention of profits might be proposed and approved by the Sovereign Wealth Fund as an alternative to a direct allocation from the Fund to the corporation.)

There is, however, another sort of issue for corporate governance in the context of this proposal. At the beginning of the twenty-first century, with some qualifications, corporate directors are formally elected by the shareholders. This proposal envisions a situation in which the Sovereign Wealth Fund would be a large plurality shareholder in most corporations. As a plurality shareholder the fund might in principle appoint the directors, or a considerable number of them. This would in effect transform the corporation into a quango. In essence this is Galbraith's proposal.[23] In the interest of decentralization and free initiative we might wish to avoid this. An alternative would be to transform the corporations into share-issuing worker cooperatives, along the lines suggested by McCain, or perhaps community cooperatives, benefit corporations, or some form yet to be developed.[24]

In practice the Sovereign Wealth Fund would not be the only institutional owner of corporate shares, nor would the corporations be the only economic actors. Retirement funds, insurance funds, and endowed nonprofit corporations would probably be owners of shares. Some residuum of individual share ownership might or might not exist. Proprietary businesses would most likely coexist with the planned sector, as well as fully member-owned cooperatives. Here, again, much must be left out in the interest of brevity. Would it be appropriate for the Sovereign Wealth Fund to accept proposals from non-profits and cooperatives for projects to be funded by loans or on other non-ownership terms? Would it be appropriate to accept such proposals from proprietary enterprises? These questions will be left unresolved.

It has been said that the Sovereign Wealth Fund would consider proposals for the foundation of new corporations, but little has been said about how they might be originated. In principle they might come from existing corporations or individuals, but these may prove to be unproductive channels. As a part of this proposal, then, we would envision yet another sort of organization. We suppose that there is in each locality a non-profit corporation, community cooperative or quango with the specific mission of promoting a high standard of living, sound environment, and full employment for the people of the locality, principally by acting as a financial intermediary and pursuing the foundation of new enterprises and the renovation of existing ones. Let us call these organizations Local Development Agencies.* In case it is a quango, the Local Development Agency would be responsible primarily to local government. In a particular planning cycle, a Local Development quango might propose the formation of a new enterprise in its district that would be financed by a transfer of investment funds from the Sovereign Wealth Fund to the Local Development Agency. The Local Development Agency would then transfer the ownership rights in the shares or debt of the new enterprise back to the Sovereign Wealth Fund as a repayment of the fund transfer.

In addition to the production of commodities for sale, there will also be government agencies engaged in the production of public and quasi-public goods. The extent to which these activities would be integrated into the planning process, and the details of that integration, will again be left unresolved.

Market economies with representative democratic governments have achieved many successes in the nineteenth and twentieth centuries, and the first three sections of this chapter are submitted as a possible explanation for those successes. The point of this section is that, if that explanation is a correct one, there may be other possibilities that could succeed for the same reasons, and it is reasonable to continue to explore these alternatives.

CHAPTER SUMMARY

A hypothetical grand coalition of the whole society might recognize the higher informational cost of cooperative decision-making and rely on non-cooperative decision-making, in the context of a market game, to better approximate efficient cooperative action. However, this arrangement falls short on a number of counts, some of which were well understood in the neoclassical welfare economics of the mid-twentieth century, and some of which have arisen from more recent research or from traditions outside the neoclassical framework. This implies a role for hierarchy,

* Of course, many community development organizations already exist. While this proposal is not represented as novel, this section proposes a qualitatively new and far more powerful function for Local Development Agencies, in the planning process. Nevertheless, the existence of such bodies within existing market economies is encouraging, since it reflects a need already perceived and to some extent addressed.

rules, and punishment, that is, government in this hypothetical grand coalition. Where elections and voting provide the means of designating the hierarchy and determining the rules and punishments, we cannot describe that arrangement as reliance on non-cooperative action. Voting is cooperative. However, representative government tolerates cooperative deviations, that is, the formation of factions and parties, a form of spontaneous cooperative action that seems to be widespread. To the extent that these factions play something like a two-stage game of bargaining and voting in the government, we may hope for a government that balances efficiency against informational costs and the equalizing tendencies of symmetrical voting opportunities. We should not assume that a market system is the only possibility for mobilizing non-cooperative action for the cooperative purposes of the grand coalition of the whole society: but, if economic planning is to have a future, it must take a form that realizes the advantages of market organization (great or slight as they may be) and improves on them.

SOURCES AND READING

(1) The first two sections of this chapter (at least) follow a trail blazed by William Baumol (1952, revised edition 1964), *Welfare Economics and the Theory of the State*, Cambridge, MA: Harvard University Press, which was distinguished from contemporary literature by treating welfare economics, not as a normative prescription, but rather as a predictive hypothesis with respect to the activities the state might be expected to undertake.

(2) Welfare economics was founded by Arthur C. Pigou (1920), *Economics of Welfare*, London: Macmillan. Pigou's approach relied on Benthamite or "cardinal" utility concepts, and a major project of economic theory in the second quarter of the twentieth century was the revision of welfare economics on the basis of what philosophers would call preference utilitarianism. While many economists contributed to this trend, Samuelson's contributions were wide-ranging and definitive. In addition to Paul A. Samuelson (1947), *Foundations of Economic Analysis*, Cambridge, MA: Harvard University Press, see Paul A. Samuelson (1948), "Consumption theory in terms of revealed preferences," *Economica*, **15**(60), 64–74; (1954), "The pure theory of public expenditure," *Review of Economics and Statistics*, **36**(4) (Nov.), 387–9; (1956), "Social indifference curves," *Quarterly Journal of Economics*, **70**(1) (Feb.), 1–27; (1977), "Reaffirming the existence of 'reasonable' Bergson-Samuelson social welfare functions," *Economica*, New Series, **44**(173) (Feb.), 81–8.

(3) The literature on the excess burden of taxation is enormous. For

a recent and magisterial discussion, see James A. Mirrlees et al. (2011), "Tax by design," Institute for Fiscal Studies, available at http://www.ifs.org.uk/mirrleesReview/design, as of 21 March 2013. (4) For Okun's ideas see Arthur Okun (1981), *Prices and Quantities*, Washington, DC: Brookings Institution. (5) Once again, the quotation from Adam Smith is from Book V, Chapter I in *The Wealth of Nations*, op. cit., Chapter 1 of this volume, note 7. (6) For Samuelson's classical contribution see Paul Samuelson, op. cit. (1954). (7) Elinor Ostrom shared the Nobel Memorial Prize in 2009 for her studies of such arrangements. See the Nobel website at http://www.nobelprize.org/nobel_prizes/economics/laureates/2009/, as of 16 November 2013. (8) I have discussed this in a more formal way in Chapter 9, section 3 of Roger A. McCain (2009), op. cit., Chapter 3 of this volume, note 17, and Chapter 9 of Roger A. McCain (2013), op. cit., Chapter 3 of this volume, note 15. (9) Reinhard Selten, op. cit., Chapter 1 of this volume, note 9. (10) George A. Akerlof and Janet Yellen (1985), op. cit., Chapter 6 of this volume, note 16. The game example in this section is influenced by the "left Austrian" model in Paul Rosenstein-Rodan (1943), "Problems of industrialization of Eastern and South-Eastern Europe," *Economic Journal*, **53**(210/211) (June–Sept.), 202–11. Keynesian economics, of course, is derived from the writing of John Maynard Keynes – see Chapter 6 of this volume, note 18. (11) Here again the interested reader may best consult the Nobel website for the prize to Akerlof, Stiglitz, and Spence for more details, at http://www.nobelprize.org/nobel_prizes/economics/laureates/2001/, as of 22 May 2013.

(12) This subsection follows particularly the ideas of Schumpeter's last writing on entrepreneurship, Joseph A. Schumpeter (1947), op. cit., Chapter 6 of this volume, note 20. (13) Psychologists have made useful studies of creativity. I have relied a good deal on Margaret A. Boden (1991), *The Creative Mind: Myths and Mechanisms*, New York: Basic Books. For a view from artificial intelligence, see Marcus Pearce and Geraint A. Wiggins (2002), "Aspects of a cognitive theory of creativity in musical composition," 2nd International Workshop on Creative Systems, European Conference on Artificial Intelligence, Lyon, France, at http://www.doc.gold.ac.uk/~mas02gw/papers/ecai02.pdf, as of 16 November 2013. Note that Schumpeter, in a recently discovered essay, used painting to explain his views on creativity: Joseph A. Schumpeter, Markus C. Becker and Thorbjørn Knudsen (2005), "Development," *Journal of Economic Literature*, **43**(1) (Mar.), 108–20.

(14) See Thorstein Veblen (1918), *The Instinct of Workmanship and the State of the Industrial Arts*, New York: B.W. Huebsch on the "instinct of workmanship." (15) The analytical literature on voting is, of course, enormous, dating as it does from the eighteenth century, but the discussion of

voting as a non-cooperative game, from which we draw here, dates from the 1970s and drew on the famous Arrow impossibility theorem: Kenneth J. Arrow (1951), *Social Choice and Individual Values*, New York: Wiley. See especially Alan Gibbard (1973), "Manipulation of voting schemes: A general result," *Econometrica*, **41**(4), 587–601; Mark A. Satterthwaite (1975), "Strategy-proofness and Arrow's conditions: Existence and correspondence theorems for voting procedures and social welfare functions," *Journal of Economic Theory*, **10**(2) (April). Allan Feldman (1979), "Manipulating voting procedures," *Economic Inquiry*, **17**(3) (July), 452–74 provides a good interpretive overview. (16) The discussion of electoral democracy in this chapter and the next will echo some themes from John Stuart Mill (1991), *Considerations on Representative Government*, New York: Prometheus Books, originally published 1861, available at http://www.gutenberg.org/ebooks/5669, as of 30 July 2013. Key points are Mill's stress on the balance of power in society as a necessary condition for electoral government and on the primary importance of deliberation, as against simply voting, in representative assemblies. (17) For a hypothetical numerical example of a voting game with multiple Nash equilibria, see Chapter 20, Table 20.4 in Roger A. McCain (2014), op. cit., Chapter 1 of this volume, note 8. On the two-stage model of negotiation with voting as a second stage, I can only refer to some of my own research: Roger A. McCain (2013), "Bargaining power and majoritarian allocations," a paper presented at the fall conference of the Atlantic Economic Society International, Philadelphia, October 2013. This applies a bargaining model first proposed in Chapter 6 of Roger A. McCain (2013), *Value Solutions in Cooperative Games*, Singapore and Hackensack, NJ: World Scientific.

(18) W. Arthur Lewis (1969), *Principles of Economic Planning*, London: Allen and Unwin. (19) Roger A. McCain (1985), "Economic planning for market economies: The optimality of planning in an economy with uncertainty and asymmetrical information," *Economic Modelling*, **2**(4) (Oct.), 317–23; (1991), "A theory of economic planning for market economies: The optimality of planning," in S. Baghwan Dahiya (ed.), *Theoretical Foundations of Development Planning*, New Delhi: Concept Books. (20) The classical statement of market socialism is Oskar Lange and Fred M. Taylor (1938), *On the Economic Theory of Socialism*, edited by Benjamin Lippincott, Minneapolis, MN: University of Minnesota Press. (21) The idea of weighted cost–benefit calculations, with weights to incorporate judgments on the distribution of income, arose from attempts to reconcile the social welfare function approach to welfare economics with the practice of cost–benefit economics. It can be found in Otto Eckstein (1961), "A survey of the theory of public expenditure criteria," in *Public*

Finances: Needs, Sources, and Utilization, Universities-National Bureau Committee for Economic Research, pp. 439–504. It was discussed and sometimes applied in the period of the 1970s but seems to have been little discussed since. (22) While the proposal of "Pigovian" taxes of course originates with Pigou, a good discussion can be found in William J. Baumol (1972), 'On taxation and the control of externalities', *American Economic Review*, **62**(3), 307–22. (23) Galbraith's "socialism" is discussed in John Kenneth Galbraith (1973), *Economics and the Public Purpose*, Boston, MA: Houghton Mifflin. (24) Roger A. McCain (1977), "On the optimum financial environment for worker-cooperatives," *Zeitschift für Nationalokonomie*, **37**(34), 355–84. Traditional cooperatives, wholly owned by their members, probably would not play a large role in the planning process discussed here, though they might be partly financed by loans from the Sovereign Wealth Fund. On community cooperatives, see Joel Magnuson (2008), *Mindful Economics*, New York: Seven Stories Press.

9. Macroeconomics

The previous chapter explored the hypothesis that the entire population of a territory might form a cooperative grand coalition. In a world of costly information, such a coalition might rely on less costly non-cooperative decision-making for many of its routine decisions. In the ideal case this might mean a market system. But this ideal case, in which markets are incentive-compatible, can be realized only if information is costless so that in practice the market decision-making can only be approximately incentive-compatible. At the same time, recognizing the role of organization in conserving costly information, the grand coalition would undoubtedly establish an organization, that is, a state. The chapter then explored some of the ways the state and the market mechanism might jointly better approximate a cooperative solution for the grand coalition than either might do separately. In the course of this exploration, a large part of twentieth-century welfare economics based on neoclassical approaches could be recovered. However, one set of issues was left for this chapter: the issues that arise if a failure of coordination among the many non-cooperative decision-makers in the economy results in a situation of deficient aggregate demand. While there will be little explicit reference to the hypothesis that government functions as an agent of the grand coalition of the whole population, this hypothesis is in the background throughout the chapter.

AGGREGATIVE ECONOMICS

In the mid-twentieth century it became common to distinguish between macroeconomics and microeconomics as subfields within economics. Roughly speaking, microeconomics was thought of as the study of equilibria in markets for particular goods and services, while macroeconomics would be the study of developments in the market economy as a whole. In fact, the growth of macroeconomics reflected two particular developments in the market economy as a whole: recession and unemployment, particularly in the period of the 1930s. By recession, of course, we mean decreases in production. Later, inflation and economic growth (or its absence, stagnation) came to be topics of macroeconomics as well. John Maynard

Keynes had offered an explanation of the recession and unemployment, together with a defense of some government policies aimed at ameliorating them. Keynes's explanation relied on a concept of equilibrium quite different from that of contemporary microeconomics. The split into two subfields allowed specialists to apply the theories that seemed most promising to the two kinds of problems. An important influence in this direction was an introductory textbook by Paul Anthony Samuelson.(1)

"The market economy as a whole" is a very complex object, and accordingly macroeconomics has relied on sweeping simplifying assumptions, even to a greater extent than microeconomics. One very important simplifying assumption is aggregation. For example, production in the economy as a whole is represented as gross domestic product: an aggregate index of what are in fact a great diversity of goods and services. Similarly investment and consumption are treated as if they were homogeneous aggregates; the labor force as a mass of undifferentiated labor resource. It is often assumed that average prices and quantities for these aggregates are determined as if by the equilibrium of a small number of aggregate markets, such as a labor market and a market for bonds. This is so much a matter of consensus that macroeconomics is often called aggregative economics.

Of course, as Einstein is misquoted, we want to make things as simple as possible, but no simpler. To the extent that aggregative economics can serve us effectively in understanding, anticipating, and making policy for real economic events, we would recognize aggregation as a brilliant insight. With one qualification, this success has yet to be shown. (The fiscal policy of the Kennedy-Johnson administration in the USA, largely designed by Professor Walter Heller, was widely considered successful in its time: this is the qualification.)

However, there is another side of the coin. Can we do without aggregation? The answer is probably not – indeed, economics is aggregative in every instance. The market for wheat is an aggregate concept and just as much an analytical fiction as the market for gross domestic product or for bonds. There is, no doubt, a difference of degree, but the issue would seem to be not whether to aggregate but what and how to aggregate.

Friedrich von Wieser writes(2) (of a somewhat different kind of economics):

> [T]his method . . . idealizes. It does not copy nature, but gives us a simplified representation of it, which is no misrepresentation, but such as sharpens our vision in view of the complexities of reality, – like the ideal picture which the geographer draws in his map, as a means not to deception but to more effective guidance, he meanwhile assuming, that they who are to profit by the map will know how to read it, i.e. to interpret it in accordance with nature.

This may be said especially of macroeconomics. Economists use various "maps," and by contrast with microeconomics, macroeconomics is a "map" on a very large scale, with relatively little detail. Economists may differ with one another about the territory that is to be mapped, but even when they agree, there may be further disagreements about the particular details that need to be included in the large-scale map of macroeconomics.

COORDINATION FAILURE

For Keynes, the key aggregates were consumption and capital formation, with capital formation taken in two senses, as saving and as investment. Keynes gave Thomas Malthus credit for having argued that "general gluts" could arise because of what twentieth-century economics would call deficient aggregate demand, but Keynes adopted the explanation of general gluts that John Stuart Mill had put forward:[3] a failure of coordination between savers and investors.

As in the previous chapters, we will illustrate the idea central to Mill and Keynes with a very oversimplified non-cooperative game example. We suppose that there are just two decision-makers: a saver/consumer and an investor. The investor can choose between two strategies, which are methods of production: simple methods and roundabout methods. As the Austrian School of Thought has taught us, the roundabout methods will be feasible only if there are ample savings to finance them. The saver/consumer can choose between greater or less saving. Payoffs are measured on a relative scale of 1–5. The game is shown in Table 9.1.

Table 9.1 A Keynes-Mill Game

First Payoff to Saver, Second Payoff to Investor		Investor	
		Simple	Roundabout
Saver/consumer	Less saving	3,3	2,1
	More saving	1,2	5,5

Here is the reasoning behind the relative payoffs: at the lower right, well-financed roundabout production is profitable even after paying an ample rate of return to savers. At the upper left, both the rate of return and the profits are less, because of the lesser productivity of the simple methods of production. At the upper right, roundabout methods largely fail because of insufficient finance, and the return to savers is reduced via bankruptcies and defaults. At the lower left, returns to savers are reduced as a result of their competition in an asset market glutted with capital, and the profits of producers are depressed by the lack of aggregate demand from consumers who, on the one hand, have little to spend and, on the other hand, have decided to save rather than spending a large proportion of that little.

Several points may be made about this game. First, it has two Nash equilibria, one where savers/consumers choose less saving and investors choose simple methods, and the other where savers/consumers choose more saving and investors choose roundabout methods. Second, the latter Nash equilibrium is Pareto-preferable to the former, and is the only strong Nash equilibrium. Thus, if there really were only two decision-makers, they undoubtedly would sit down together and agree on the strong Nash equilibrium. But here, as usual, the two-person game is an extreme simplifying assumption. In the actual world of many millions of decision-makers and costly information, this multilateral conference will not be feasible, and we cannot assume that the strong Nash equilibrium will be the one realized. Some further information will be required to assure that.(4)

Third, because there are two Nash equilibria, all four strategies are rationalizable. Thus, to the extent that decision-makers rely on rationalization, we cannot even be certain that a Nash equilibrium will be observed. Suppose, for example, that saver/consumers reason as follows: "Investors believe that consumer confidence is very low, so that we will expect low incomes and returns and so commit ourselves to less saving. Their best response to that is to choose simple methods, and that means my best response to that is to choose less saving, as they expect." But suppose the consumers are mistaken about investors' expectations and the investors reason: "Consumer confidence is very high, that is, consumers expect us to choose roundabout methods with resulting high incomes and high rates of return to savers. Their best response to that expected behavior is to choose more saving, and our best response, in turn, is to choose the roundabout methods they expect." But these rationalizations would lead to the non-equilibrium outcome at the upper right – a crisis with many business failures and consequent declines in income and returns. But both kinds of decision-makers have been mistaken, of course, so let us further suppose that each corrects their errors in the next period, so that savers/consumers reason: "Investors have shown that they believe consumer confidence is high, so that we will choose more saving. Their best response is to choose roundabout methods, as they have done. Accordingly, our best response is to choose more saving, as they expect." The investors, however, will be thinking: "Consumers have shown that their confidence is very low, that is, consumers expect us to choose simple methods with resulting low incomes and low rates of return to savers. Their best response to that expected behavior is to choose less saving, and our best response, in turn, is to choose the simple methods they expect." We then find ourselves in the non-equilibrium situation at the lower left – a recession with a capital glut. And while this example is simplified near the point of parody, it does bear some relation to the sequences of events observed in the USA during 1929–31 and 2007–09.

Fourth and finally, rationality per se – that is, rationalization – is not sufficient to determine the optimal policies or the outcome. Perhaps this is what Keynes meant when he said that decisions cannot be rational but "can only be taken as a result of animal spirits" so that "if the animal spirits are dimmed and spontaneous optimism falters, leaving us with nothing but a mathematical expectation, enterprise will fade and die."(3) (Keynes actually makes reference to the Keynesian beauty contest, a pure coordination game, as an explanation of observed wide fluctuations in security markets. However, the coordination game shown as Table 8.1 in Chapter 8 suggests another way in which investment may depend on "the state of business expectations.")

Nevertheless, the Keynes-Mill Game departs from Keynes's thinking in one important way. In the game, we see recession-like phenomena at the disequilibrium outcomes. The equilibrium in the upper left is not a "general glut" so much as it is a "general dearth" or a low-level economic development trap. Keynes assumed that saving decisions would be conditioned *primarily* on current income, and from that assumption, he derived a model with a unique equilibrium much more in the style of economic theory at that time. Much ink was subsequently spilled – and some very important new things learned – in mid-century economic research that showed that Keynes was mistaken in believing that saving would in general be conditioned on current income.(5) But this is an epic case of missing the point. If indeed, as the game suggests, multiple rational-action equilibria can occur in the coordination of saving and investment decisions, then non-cooperative rationality fails much as we have seen, regardless of the relation of saving to current income. For the economics of mid-century, however, models with multiple equilibria simply were inadmissible on methodological grounds. It was only after Schelling's work(6) in game theory that we began to realize that multiple non-cooperative equilibria could be an explanatory principle, rather than a symptom that too few assumptions had been made. Keynes did his best to reconcile his theory with the methodological principle that only equilibrium models with unique equilibria are admissible.

As we have said, Keynes's aggregative economics was based on a very different equilibrium concept than market clearing equilibrium, but, nevertheless, in many ways they are quite close. In the market equilibrium approach to economics, the only cooperative elements are bilateral contracts. These contracts are arrived at by some sort of non-cooperative interaction. Following Walras, market equilibrium theorists assume that prices are determined "as if" by non-cooperative bidding in an auction game. Keynesian equilibria are also non-cooperative. The differences are in the strategies among which the players may choose and the rules of the game that are included in the model or the "map" and those that are abstracted from.

Despite its predominant role in economics *teaching* in the latter half of the twentieth century, Keynesian macroeconomics was never uniformly accepted by the economics profession. The anti-Keynesian position is that the payoffs in the Keynesian investment game, and Keynes's assumed relation between consumption and investment, would not be stable in the face of changing market prices. On this view, whenever investment is less than the efficient amount, or whenever there is unemployment, market prices would change so as to eliminate the Nash equilibria at less than full employment. But the obvious difficulty of this view is that we do observe some unemployment and it is quite persistent in some periods and fluctuates with production in roughly the way Keynesian economics predicts. To the extent that this is true, there must be at least some markets that do not clear. From the writings of Keynes's rival (and mentor) Arthur C. Pigou to the present, most critics of Keynesian economics have focused on labor markets as the prime suspect. Some of the work done on macroeconomics and labor markets in the late twentieth century was honored by the Nobel Memorial Prize of 2010.(7)

In the next section some of the ideas common to most macroeconomic research in the late twentieth century will be sketched, and in the ones to follow, the Nobel Laureate work on labor markets will be reconsidered from the point of view of imperfect cooperation.

DYNAMIC STOCHASTIC GENERAL EQUILIBRIUM

Most contemporary macroeconomic models, including those described as "New Keynesian," apply what is called the dynamic stochastic general equilibrium, DSGE, approach. This is a pretty large family of theories, but what they have in common is captured by those four hifalutin' words: "dynamic stochastic general equilibrium." "Dynamic" in means in part that the equilibrium we study is not stationary. Rather, it is a moving equilibrium that determines some constants and persistent trends on the basis of long-run influences. "Stochastic" means that the economy is disturbed away from the moving equilibrium from time to time by unpredictable shocks. "Equilibrium" means that the long-run influences and the responses to the shocks are determined by the rational decisions of human decision-makers, perhaps through the equality of supply and demand or perhaps through more complex models that allow for deviations such as monopoly power, costs of transaction and adjustment, frictions and rigidities in key markets, and the bounded rationality of real human beings. It is here that different theories in the DSGE family vary most. Finally, "general" in "dynamic stochastic general equilibrium" means that equilibria in this sense exist at the same time in all *key* markets, such as labor markets, financial markets, and markets for capital goods, and (on the average, at least) in markets for finished goods and services.

The idea of moving equilibrium was an important step forward for neoclassical economics at the mid-twentieth century. Neither the timeless world of Walrasian market equilibrium theory nor the stationary state of classical political economy seems an adequate representation of our world. The moving equilibrium reflects, on the one hand, growth in the resources available to the economy, and, on the other hand, technical progress. The resources that grow are, on the one hand, labor, and on the other hand, the supply of capital goods such as machines, computers, and factory buildings, and so forth, as well as human capital. (This is, of course, a highly simplified view.) But growth, per se, does not make people better off. If the labor force grows, but other resources do not, and the technology does not improve, then (as Malthus taught) people become worse off, due to the diminishing marginal productivity of labor. This can be offset by investment, which creates more resources per worker; but this, too, is subject to diminishing marginal productivity, so that (as Malthus and the classical political economists believed) the economy could settle into a stationary state. This would occur if the technology did not improve. Conversely, the moving equilibrium of the economic system is one in which the improvement of technology just offsets the diminishing marginal productivity of labor and capital, so that production per worker increases steadily.[(8)]

If improvement of technology were to bring about a steady growth of the productivity of labor (with given investment per worker) and some other influences on the economy are held steady, the model would predict a steady growth of gross domestic product and of wages, without recessions and unemployment. However, even if it is predictable on the average over a long period of time, labor productivity improvement is very unpredictable in any particular year or shorter period. This random variation in the rate of productivity growth is one example of the random shocks that would disturb the economy and shift it away from its moving equilibrium. Some macroeconomic models allow only for these productivity shocks, but for no other shocks or deviations from market clearing equilibrium.

These are called real business cycle models. Other models allow for other kinds of shocks and deviations from market clearing equilibrium as well.

Once the economy has been displaced from its moving equilibrium, the *people* whose decisions drive the economy will respond to the conditions, causing further changes that further modify the path of the economy. These further changes can be thought of, in the same dynamic spirit, as feedbacks. These feedbacks determine the short-term trends of the economy. If government attempts to stabilize the economy, this is yet another feedback. It is here that economists may disagree most on the territory that is to be mapped – which feedbacks are more important and which are less so? If feedbacks move the economy toward the moving equilibrium, they are negative feedbacks, while feedbacks that move the economy even further away from the moving equilibrium are positive feedbacks. In dynamics, negative is good and positive is bad (usually).

The key differences between Keynesian and anti-Keynesian economists have to do with different beliefs about the feedbacks. Anti-Keynesian economists tend to focus on feedbacks via labor markets, which are supposed to be negative feedbacks; and they argue that the feedback via government policy is positive, rather than negative. In the words that would probably be used, their position is that government policy is destabilizing. Keynes and Keynesian economists have focused instead on feedbacks from expenditure to income. That is, in response to a shock, some people may cut back their expenditure, but this reduces the incomes of other people who then cut their own expenditure, moving the economy further away from the moving equilibrium, in the direction of recession. In response to other shocks, some people may increase their spending, increasing the income of others and causing increases in their expenditure, moving the economy further from its moving equilibrium in an inflationary direction. Keynesian economists would argue that government policies aimed at offsetting the positive feedbacks could improve the performance of the economy in some circumstances. Indeed, government fiscal policy and monetary policy are often referred to as stabilization policy by economists in the Keynesian tradition.

Now, Keynes believed that the positive feedback from expenditure to income would be limited. The limit of this positive feedback would correspond to a Keynesian equilibrium that might differ from the (full employment) moving equilibrium with clearing of all markets. We have seen(5) that some studies of consumption in the 1950s imply that the positive feedbacks would be less powerful than Keynes supposed. On the other hand, there may be other positive feedbacks that are sometimes far more important, as when the bankruptcy of some businesses results in impairment of the financial soundness and resulting bankruptcy of still other companies.

If the negative feedbacks in labor markets were powerful enough to overwhelm all the others, then the anti-Keynesian position would seem correct, but it is not clear that this is so. It seems that the dynamics of a modern market economy are more complex than envisioned by either Keynesian or anti-Keynesian economics. Moreover, there is reason to believe the labor market feedbacks are quite limited.

LABOR MARKETS: NON-COOPERATIVE APPROACH

Unemployment is one of the most evident characteristics of major economic downturns such as those of the 1930s, 1980s, and 2000s. No doubt that is the reason why economists often focus most on labor markets in their attempts to understand business fluctuations. And there is something intrinsically puzzling about unemployment. By definition, a person is unemployed only if she or he is willing to work at a profitable wage but has no job. On its face, it would seem that every unemployed potential worker is a missed opportunity for profit. In a capitalist economy, we do not expect opportunities for profit to remain unrealized for very long. Yet unemployment can be quite persistent. How can we account for that?

To do that we will have to shift our understanding of equilibrium in a small but important way. In real business cycle models equilibrium is taken to mean that markets clear: quantity supplied is equal to quantity demanded. If there is involuntary unemployment then the labor market does not clear. Instead, we will understand "equilibrium" to mean that every decision-maker in the economy makes decisions consistent with her rational best interests in a non-cooperative sense. If markets clear, then we can be sure that that is so – clearing markets are a sufficient condition for equilibrium. But one thing we have learned from game theory is that clearing markets are not a necessary condition for equilibrium. Non-cooperative game theory gives us many examples of equilibrium in which markets are not defined, let alone do they clear. Thus, we will be looking for an equilibrium in that broader sense, a Nash or similar non-cooperative equilibrium.

(In this section we will adopt the neoclassical understanding of "rational" as non-cooperative rationality and that the only cooperative phenomena are bilateral contracts. These are standard assumptions in the Nobel Laureate research we are discussing. These assumptions will be reconsidered in the next section.)

There are a number of hypotheses in recent economic research that attempt to account for the persistence of unemployment. (By recent I mean roughly the last 30 years.) One possibility is transaction costs, and specifically costs of hiring or firing. For now we will focus on hiring costs. There is good reason to think that hiring has a cost: in order to fill a job, it is necessary to match the employer's vacancy and its requirements with the employee's availability and capabilities. This matching will take time and resources. The cost will be higher if fewer potential employees are available, and the smaller the number of unemployed the fewer are available.

This is not to say that only unemployed workers are available. The vacancy may be filled by an employee who has come directly from another job. But (1) that is likely to lead to yet another hiring, to replace the employee who has moved on, and (2) a much higher proportion of unemployed than of employed workers will be available for a particular opening. Thus, some studies make the simplifying assumption that available workers and unemployed workers are identical. For this section we will adopt that simplifying assumption.

The assumption of costly search and matching can easily account for the existence and persistence of unemployment. Suppose that the labor force includes N workers and H are employed, with $H < N$. Unemployment is $(N - H)/N$. Suppose that the value of the marginal product of labor is V; the marginal disutility of labor, expressed in money terms is U; $V > U$, and C is the cost of matching a worker to a job. In other words, V is the highest wage that any firm could pay without reducing its profits, and U is the lowest wage that anyone not employed could accept without being worse off. If $C = 0$, as in neoclassical economics, this is an unstable situation: someone will be hired for a shared net benefit of $V - U$ and unemployment will decrease. Suppose, instead, that $C > V - U$. If a firm hires an unemployed worker under these circumstances, the firm will lose at least $U + C - V$, and so there will be no hiring and unemployment will be stable at $(N - H)/N$.

There are several complications that need to be discussed:

1. The cost of hiring will be shared between the employer and the candidate employee. Thus, $C = C_f + C_w$, where C_f is the cost borne by the firm and C_w the cost borne by the candidate employee. If the expected wage is W and $U + C_w > W > U$, the candidate cannot expect a position that would repay both his disutility and the cost of seeking the position, and so the person will not seek employment but instead is a discouraged worker. In part, C_w may be the cost of waiting in the unemployment queue, especially if the candidate refuses an offer in order to wait for a better match. In part, however, the candidate will have to commit some resources to the search, such as the cost of travel to an interview.

2. Once the match is made and the worker employed, C_f and C_w are sunk costs, and thus not further relevant to decisions within the firm.
3. The situation is both better and worse for the employer than our discussion so far suggests. On the one hand, the employment relation will typically continue for some time. After the employee has been hired, the ongoing relation of employer and employee will produce a surplus in every period that the worker remains employed. On the other hand, employees will quit (or otherwise leave the employment relation) from time to time, so that the employer will have to bear recruitment costs even to maintain the firm's work force. Thus, we must interpret V and U as discounted values of flows of the expected value of value product and utility over a foreseeable term of employment. This creates some problems for computation, and it seems fair to say that they have not been solved, since models that address this issue tend to make simplifying assumptions that are themselves problematic. It is common to assume that there is a fixed and given probability that the employment relation will terminate in any given period. For a rational action theory this probability should depend on the motivations of the employer and employee – even if their rationality is bounded.
4. The value of the marginal product can only be an expected value of an unknown number. If it were known with certainty there would be no need for search and thus no search cost. Instead, at least one component of the value of the marginal product of a *particular* employee in a *particular* job will depend on the *particular* match between that job and that employee. This component of productivity is called an idiosyncratic shock to productivity. What the search process does is produce information about the value of that idiosyncratic shock. Of course, the firm will do what it can (at low cost) to narrow the field by providing information about its own needs: for example, if the position requires engineering skills, it will consider only applications from qualified engineers. This is why we have many labor markets rather than one big one. Nevertheless, it is the individual match that matters for the idiosyncratic shock, since otherwise costly search would not be justified.
5. One of the advantages of the search cost approach is that it allows us to relate unemployment to benefits and costs, and thus provides at least a tentative basis to judge the efficiency of policies to increase (or reduce?) employment. Suppose that $V - U > C$. Then it will be profitable for some firm to hire some unemployed worker, reducing unemployment. This will produce changes in the values of V, U, C_f, and C_w. In the neoclassical tradition we might expect V to decrease (due to decreasing marginal returns in the short run) and U might

increase if the remaining unemployed workers have higher disutilities of labor. But C_f and C_w will also change.

The additional hire will reduce the number of unemployed, leading to tighter labor markets. This will tend to increase C_f, in that there will be fewer candidates for each remaining opening. It will at the same time reduce C_w, in that with tighter markets, the wait for an acceptable offer and the number of interviews one must take to find it will be less on the average, there being fewer rivals in the market. Moreover, these cost changes are (mostly) externalities, since C_f and C_w are averages over all firms and all employment candidates respectively. In treating them as externalities we are applying the large numbers assumption: the firm's hiring will be so small, relative to the pool of available workers, that it will have no (perceptible) impact on the cost per hire; and similarly the decision of an individual employee candidate among the very many potential candidates will have no perceptible impact on hiring costs. But even though the change is so small relative to the labor market as a whole that it does not influence the decisions of the employer and the candidate, the aggregate of these decisions by many small employers and many candidates will give rise to noticeably tighter labor markets.

Thus, an increase in employment will generate both positive and negative externalities. The efficient level of unemployment will be one that just balances the positive and negative externalities along with the other benefits and costs of employment. But, as always in the presence of externalities, there is no reason to think that the market will realize an efficient rate of unemployment, and it would be a great coincidence if it did. There is some reason to suspect that when unemployment is high, increases in employment generate net positive externalities, while when unemployment is very low, the reverse is so.

6. In neoclassical economics, the wage is determined by the equilibrium of supply and demand. In the presence of search costs, however, the wage is indeterminate within some limits. Consider a particular firm, and suppose it pays its employees a wage W, with $U + C_w + C_f > W > U + C_w$. In the absence of employer hiring costs, labor would be available at a market wage of $U + C_w$, so the firm could discharge its existing employees and replace them at a lower wage $W^* = U + C_w$. In the presence of employer hiring costs, however, the cost of replacement workers would be $U + C_w + C_f$, and profits would be reduced. Similarly, an employee might observe that jobs are available at other employers at a wage $W^\dagger$ with $W + C_w > W^\dagger > W$. In the absence of employee search costs, a rational employee would quit to seek a job at the alternate employer. But with $C_w > 0$, this

would leave the employee worse off. Thus, so long as $U + C_w + C_f > W$ and $W + C_w > W^\dagger$, we have equilibrium, in the sense that neither the employer nor the employee can benefit by changing their decisions among the alternatives available to them. Thus, (1) for a particular firm the wage is indeterminate within these limits and (2) we should expect to observe different wages in different firms, in some cases.

7. Once again, V is an average over all firms in the labor market. Suppose a particular firm experiences a firm-specific shock that reduces its value of marginal product V^* so that $V^* < W$, the wage it has agreed to pay. Then the firm will find it profitable to lay off or dismiss* some or all of its work force, not to replace them but to "rightsize" the firm. In the absence of search costs, this would be efficient, as the employees would simply move on to other employment at their best alternative opportunities. Recall, however, the C_f and C_w are sunk costs. So long as $V > U$, the employment relation continues to generate a surplus, so its continuation is efficient. This might require a renegotiation of the wage, but such a renegotiation could require a degree of trust between employer and employee that is missing, especially if the contract between them is implicit (Chapter 6). In any case, the dismissal, if it occurs, will increase overall unemployment, generating negative externalities (to unemployed workers) and positive ones (to other employers). If the negative externalities are predominant, this could justify what Keynesian economics calls stabilization policy.

The fact that the wage is indeterminate in the presence of search costs is a recognized problem for the theory. Some authors have attempted to solve the problem by applying bargaining theory, often said to be "non-cooperative" bargaining theory. These models have not been very successful. There are three difficulties. First, and now widely recognized, they do not fit with empirical evidence. Second, as Nash observes in his seldom read 1953 paper on cooperative games, there are infinitely many Nash equilibria in a simple bargaining game, so Nash equilibrium per se simply cannot answer the question. Finally, Nash's axiomatic bargaining theory requires equal bargaining power, while the evidence suggests that bargaining power is often unequal and varies widely.(9) Thus, it is not to be expected that non-cooperative rationality can explain the sharing of the surplus among the parties to an agreement.

* So far as I know, there is no research in the search theory tradition that addresses the difference between layoffs and dismissals and the reasons why firms often choose layoffs rather than dismissals.

MATCHES FOR COOPERATION

As we have seen in previous chapters, there are several reasons why non-cooperative decision-making would take place in a population in which decisions are *essentially* cooperative. One reason put forward in Chapter 7 is that cooperative decisions cannot be made until a cooperative coalition has been formed, so that the decision to form a coalition must unavoidably be non-cooperative. The models of job matching honored by the 2010 Nobel seem to fit that pattern. The matching process is unavoidably non-cooperative, since there are no pre-existing coalitions that include both the job seekers and the employers. However, once the employment link is made, it forms a coalition that includes both the employer and the employee: the relation they enter into is one intended to realize mutual benefit, a cooperative coalition. Moreover, in many cases, the match is made between an existing coalition, comprising the employer and a number of prior employees (and shareholders and others), and the newly hired. This implies a number of complications not dealt with in either received cooperative game theory or most macroeconomic theory.

Bargaining Power of Groups

First, the bargaining models in recent macroeconomics assume that bargaining takes place at the time of hiring and is strictly two-party bargaining between a proprietor and an individual employee. This is consistent with the neoclassical supposition that bilateral contracts are the only cooperative phenomena in a market economy. However, common sense suggests that a group within a coalition may increase their bargaining power by leaguing together and threatening coordinated action. The most familiar instance of this is the strike, but even in the absence of strikes and unions, it seems likely that the work force as a whole, or strategic subgroups within it, may be able to exercise group bargaining power. For example, a coordinated withdrawal of effort could be a credible threat, especially if the hiring cost of replacing a group of workers would increase more than in proportion to the size of the group. However, the potentiality for group threats, and thus the bargaining power of subgroups within the coalition such as the work force, seem to depend on organizational dimensions of the firm that are themselves relatively permanent. It would follow that the distribution of bargaining power among groups within the firm would also be relatively permanent; and that could account for the observation that real wages do not fluctuate with unemployment – that real wages are sticky.

The withdrawal of effort does play a role in some macroeconomic models based on an efficiency wage theory.[10] We can illustrate the ideas behind this theory (with the usual caveat about oversimplification) with a two-person game, an effort dilemma. For this game, the employee can choose between two strategies: making an efficient effort ("work") or withdrawing it ("shirk"). The employer can pay the market wage rate or can pay more generously. In Table 9.2, relative payoffs are shown on a scale of 5. We see that this is similar to the Prisoner's Dilemma and the Taking Game: there are dominant strategies for both participants – pay the market rate and shirk – and there is a contrasting cooperative solution, pay generously and work. In this simplified example the cooperative strategies create value that doubles the non-cooperative values.

Table 9.2 An effort dilemma

First Payoff to Employer, Second Payoff to Employee		Employee	
		Work	Shirk
Employer	Pay generously	4,4	1,5
	Pay market rate	5,1	2,2

The efficiency wage models generally treat the strategies as continuously variable, so that effort can take a numerical value from some interval and the wage can take any value from the market rate upward. Profits to the employer are assumed to vary smoothly with effort and decisions are supposed to be made non-cooperatively, so that the employer maximizes profits and the employee balances the threat of dismissal against the disutility of effort. The solution is found by methods that resemble the market equilibrium models, though the result is that markets do not clear. Even within the non-cooperative framework, however, and focusing on the simple two-strategy game above, it is not hard to understand how the cooperative strategies might be realized. The effort dilemma is a game played repeatedly, with a large probability of repetition but not with certainty. It is well understood in non-cooperative game theory (as discussed in Chapter 7 and applied in Chapter 8 above) that the cooperative outcome can often be realized in such a case. This repeated-play interpretation seems a reasonable description (within the limits of non-cooperative rationality) of labor relations in successful firms.

From a viewpoint of cooperative rationality, however, only the upper left strategy pair can reasonably be chosen. The value created, 8 in the example, will be divided between the employer and the employee according to bargaining power. Bargaining power arises from threats. Effort withdrawal is a credible threat in either a cooperative or a non-cooperative sense. Presumably the threat to cut the wage is also credible in both senses. It is hard to say how these threats will balance out and be reflected in bargaining power, but there is no reason to suppose that the employees will have zero bargaining power. However, that is exactly what the profit-maximization model supposes.

Here the simplification of treating this game as a two-person game is again potentially confusing. As we have already suggested, a coordinated withdrawal of effort by a large proportion, or the whole, of the work force could be a more powerful threat than any threat available to an individual. This poses a problem of coordination among the employees: some signal will be needed to enable them to coordinate their decisions. In the absence of any employees' organization, formal or informal, a decision by the employer to cut the wage could provide that signal. Thus, while the non-cooperative efficiency wage models do not seem satisfactory, their central insight – that employers may pay wages above the market rate in order to avoid the threat of effort withdrawal – is an important insight about the determinants of bargaining power within the firm and so about the determination of the real wage.

Customers

Thus far we have ignored a group quite crucial to every business: the customers. Indeed the Nobel Laureate research on job matching tends to adopt either the market-clearing or the monopolistically competitive model of sales and pricing. However, these non-cooperative models of price formation are incoherent in a world of positive transaction costs. As Hall(11) has observed in a recent paper, recruitment of new customers is costly, as is the recruitment of new employees. Indeed, companies spend large amounts on advertising and other strategies to recruit more customers. Customers also invest some of their resources in finding sellers whose offers fit their particular preferences: this is called shopping. But Hall follows the crowd by assuming that, once they have recruited their customers, sellers set a single price non-cooperatively. This price must be high enough to cover the cost of recruiting customers, and thus above the competitive price that would prevail in the absence of transaction costs. The customers respond by reducing their purchases below the efficient rate, generating the inefficient result that is familiar from monopoly theory. But, even in a non-cooperative framework, why would businesses – having spent resources to recruit a group of customers – choose a pricing strategy that would result in an inefficient restriction of sales?

Second-degree price discrimination could increase the profits of the seller. Second-degree price discrimination is the charging of different prices according to quantity bought, does not require much costly information, and is as common as dirt in real-world pricing. Rationality, whether non-cooperative or cooperative, would predict some price discrimination in this situation.

However, customers are members of the production-and-sale coalition along with the proprietors and employees. Now, it must be said that there are qualitative differences in the membership of customers, proprietors, and employees, and this phenomenon – qualitatively different forms of membership for members of different types – has not been explored in formal cooperative game theory. The differences may be tied to costs of information – of course! Employees are engaged with their employer for a large part of their day, day after day, while customers are not; thus, the employee's information about the employment relation will be much richer than the customer's information about his relationship to the company. There are also differences in the impact of company decisions on the interests of the different types of agents. A customer is likely to be, at the same time, a member of several different production-and-sales coalitions for many of the same items, so that switching from one to another is an easy response to adverse decisions by the coalition. Employees, even if they have more than one job, have no such flexibility. Wage decisions by an employer have a much greater impact on the interest of the employee than price decisions do on the interests of the individual customer. Thus, bargaining is a more important determinant of the wage than of the price policy of the seller, and bargaining between the customer and the seller is likely to be implicit or episodic. On the other hand, the second-best alternative of the customer is likely to have more influence on the price strategy than the employee's second-best alternative has on the wage policy.

There will also be type differences within the group of customers. Many customer relationships are quite persistent over time, but others are not. As with the employment relationship, the affiliation of a customer to a seller produces a surplus in each period, so that customer relationships that are more persistent over time will be more valuable to the seller. Conversely, third-degree price discrimination between loyal customers and idiosyncratic ones is to be expected, and indeed is quite common.[(11)]

In passing, there are also distinctions of type among employees and, crucially, among proprietors. In particular, an active managing proprietor – an entrepreneur in Menger's sense – will have a rich information set about the coalition, while absentee owners will not. Their interests will also be differently affected, in many cases. This has already been discussed in Chapter 7.

"Business Cycles"

This has been a rather longwinded digression from macroeconomics into the theory of the firm, a branch of microeconomics. However, the issues that motivate it are macroeconomic issues. It does seem that the economics of matching cost have provided the long-sought microfoundations for a wide range of macroeconomic models. In so doing, however, this approach has brought to the fore the essentially cooperative nature of the business firm, and raised the associated issue of the determination of wages and consumer benefits by bargaining power.

The macroeconomic issue we face is the persistence of unemployment. For neoclassical economics the *occurrence* of unemployment is no mystery: it is an excess supply of labor. The mystery is why unemployment, once it occurs, is sometimes quite persistent. Keynes had attempted to explain the persistence of unemployment by a persistent deficiency of aggregate demand. The persistence of deficient aggregate demand, in turn, was explained by the equilibrium mechanism described in Chapters III and X of *The General Theory of Employment, Interest, and Money*, and in so many undergraduate textbooks.(3) However, Keynes frankly described it as a model of "temporary equilibrium" and it failed to persuade many economists as an explanation of the persistence of unemployment. It now seems that Keynes had it backward. The persistence of deficient aggregate demand is not a cause but a result of the persistence of unemployment. In turn the cost of job matching explains the persistence of a large "reserve army of the unemployed." What is recklessly called the "business cycle" often seems to go as follows:

1. A crisis of some nature destroys a number of existing production-and-sale coalitions, and/or forces others to reduce the scale of their operation.
2. The destruction of some coalitions, and the reorganization of others at reduced scale, implies a reduction of both aggregate demand *and supply*. The constructs of *independent* functions of aggregate demand and supply seems to be a fallacy of reification, as many critics have pointed out in the past.(12)
3. In the new situation, new production-and-sale coalitions are formed, and existing ones gradually recruit new members and approach their optimal scale.

This process of organization formation takes place at *about* the same rate as it did before the crisis. It is often said that there is a period of "recovery" during which the organizational economy expands at an increased rate and approaches the growth path it had been on before the crisis. This is an expression of faith and there is no evidence for it. A hypothesis no less consistent with the evidence is that the economy emerges from a crisis on a new growth path that is permanently lower than the one the economy had been before the crisis.

Interim Summary

In an economy in which employment relations are predominantly casual associations in day-labor markets and most retail sales are one-off sales in the bazaar, that is, the kind of economy modeled by neoclassical economics, the formation of organizations arguably would exercise little influence over production and employment. However, in an economy in which most production and sales are mediated by coalitions that we may recognize as organizations, the formation of organizations, a costly process, and the destruction of organizations may well be the most important determinants of shifts both in what introductory economics would call the aggregate supply and demand relationships. To the extent that they are not independent, these constructs miss the point, and we are better advised instead to address the dynamic shifts that underlie both of them together. In particular, realizing that employment decisions reflect hiring costs no less than the value marginal productivity of labor, and that wages reflect the bargaining power of insiders more than they reflect excess demands or supplies of labor, we cannot expect the negative feedbacks through labor markets to be very powerful. And conversely, in a period of crisis when the destruction of some production-and-sale coalitions may result in the failure of others, we must suppose that positive feedbacks may be quite important. All in all, the idea of the market economy as a self-stabilizing system needs to be examined carefully – and rejected!

CREATIVE DESTRUCTION

Failure and discontinuation of business firms (and other organizations) is a fact of life. Organizations are not forever. Of course, such failures are particularly to be expected in periods of recession. This is often said to be a consequence of "creative destruction," a term that comes to us from Schumpeter.[13] Here we will address three questions about creative destruction. (1) How can destruction be "creative?" (2) Does creative destruction actually occur, and is it common? (3) What, if anything, does "creative destruction" have to do with recessions?

Suppose that a new firm offers a new product, adopts a new technology, or is better organized, an innovation in the sense of the Chapter 7. Thus, the new firm is able to make offers to the customers of existing firms that those customers find preferable to the offers made by the existing firms. The existing firms, having lost their customers, cannot continue – they are destroyed. Since this occurs as a result of the innovation, a creative act, we may reasonably call it creative destruction. It will not be difficult to provide plausible examples of this from recent economic history,

Common discussions of creative destruction seem to assume that *whenever* a firm fails because of new competition, this destruction must be creative. (This is rarely said explicitly, but seems implicit in much journalistic and some scholarly writing and discussion.) This is not so. For Schumpeter, innovations occur against the background of a stationary state, a physiocratic circular flow. Schumpeter *assumes* that this circular flow can be disturbed *only* by innovation. In cooperative game theory, we have a concept that corresponds to the stationary state. It is the core of a cooperative game, a concept discussed in Chapter 3. For a stationary state to exist, it would have to be in the core of the game corresponding to the market economy. It need not be unique, for Schumpeter's purposes, and indeed for many example games, the core is not unique. But there are also examples in which the core is a null set, as we saw in Chapter 3.

Consider, in particular, the example in Table 3.8, Chapter 3, and assume that the four technologies assumed there are known from the beginning. In that example, the replacement of a singleton coalition or a doubleton coalition by a larger coalition will increase the average payout or productivity in the game, and so might be considered creative destruction of the smaller. But in these cases there is no innovation, but only the implementation of already known possibilities. Similarly the replacement of a 3/1 partition by the grand coalition could be seen as an improvement. However, the competitive replacement of the grand coalition by a 3/1 coalition, which is also to be expected in this game, reduces the average payout in this game from 4.5 to 4. This is simply destructive destruction. What the empty core model shows us is that there needs be no stationary state, and no correlation between the competitive destruction of established firms and increasing economic welfare or creative innovation.

Of course, the market economy is far more complex than this example. We have numerous proofs that market equilibria exist and are unique. But these proofs simply substitute other assumptions, and often even less plausible ones, for Schumpeter's assumption of the circular flow. Nevertheless, it seems likely that *in times of prosperity*, creative destruction is often observed. It is not, however, the only reason why firms fail. Systematic shocks to aggregate demand, labor productivity, or the terms of trade, as well as destructive destruction, can create conditions in which firms cannot survive although they could have survived had the shocks not occurred. Such systematic shocks are associated with recessions, either as cause or effect. Thus, we have no good reason to suppose that creative destruction is any more frequent in times of recession than in other times, or plays any causal role in recessions.

STAGNATION

In everything so far we have assumed, along with the consensus of modern economics, that in good circumstances the market economy will realize its potential, utilizing labor at a rate that corresponds to a stable, low, and sustainable (if perhaps not efficient) rate of unemployment. Presumably this would correspond to "expand a lot" in the coordination game in Chapter 8, Table 8.1, and to the strong Nash equilibrium in the Keynes-Mill Game in this chapter at Table 9.1. Thus, coordination problems being solved, through some combination of market processes and public policy, in periods between major shocks the economy might move toward what we think of as full employment.

If a major role of the capitalist class is to search for profitable, innovative opportunities to reinvest its profits, then a decline in the availability of these opportunities would transform the capitalist process and might itself be a generator of crises. This decline in the availability of investment opportunities is denoted as secular stagnation.(14)

As a first step, let us idealize the capitalist entrepreneurial process as follows: spontaneous technical progress provides a flow of entrepreneurial opportunities that arrive in each period according to some random arrival process such as a Poisson process. Each opportunity is characterized by an internal rate of return (in the given circumstances of the period). Entrepreneurs search among these (at some resource cost) to discover opportunities characterized by internal rates of return that exceed some aspirational profit rate. Those that do exceed the aspirational rate are funded and the sum of those funds constitutes investment for the period. If the parameters of the process that generates opportunities are appropriately stable relative to the size of the economy, the result could be a process of steady growth in which the proportion of investment to income and the aspirational profit rate (which we may then call the rate of interest) remain approximately constant. Because of the random nature of the process that generates opportunities, there will be short-term fluctuations, which might be modeled as productivity shocks, but the long-term average growth rate would be stable and depend on the parameters of the opportunity-generating mechanism.

Before proceeding it is important to notice some important criticisms that might be directed at this idealization. It is, of course, intuitive, not mathematical, but without doubt there are many economists who could generate a mathematical model to correspond to it – probably several such models. But it might be said to misrepresent the role of the entrepreneur. As Schumpeter (1947) says,[(14)] entrepreneurship is "the creative response": in a sense the innovations that drive the capitalist economy are not merely discovered by entrepreneurs but are created by them. And when we consider the creation of organizations and market channels to bring the innovation to the market economy, that is quite true; and moreover the distinction between the innovation itself and the organizations and market channels is a very blurry one, especially where the "new combination" is a service activity. But this criticism accomplishes nothing. Every decision-maker, however creative, faces constraints of limited opportunity. The creative artists Schumpeter (2005)[(14)] discussed did indeed create the artistic sensibility that came to regard their work as excellent, but they were constrained by the materials available in their time, such as oil paints and marble, no less than were their less creative contemporaries. Entrepreneurial creativity is not a magic wand that sweeps away all material limitations. The flow of opportunities the previous paragraph refers to is a flow of opportunities for *creative* search and profit-seeking; we need not distinguish the creative aspects of entrepreneurial search from the more routine ones.

Accordingly we return to the idealized process from the previous paragraph, with one change. Suppose the parameters of the random arrival process of opportunities shift over time in such a way that, at a particular aspirational profit rate, investment increases at a decreasing rate or actually declines. This, then, would be secular stagnation.

An old-fashioned Keynesian view of the market economy would hold that total production is determined as the product of the sum of four categories of autonomous spending times a multiplier that depends on the saving behavior of consumer households. The four categories are investment, net exports, government purchases of goods and services, and a component of consumption spending that is independent of income (and in the long run may be very small or zero). Taxes also play an indirect role: the higher the tax rate the smaller the multiplier is, in any given circumstances.

To do its job this multiplier must be positive and substantially more than one. As we have seen, Keynes's argument for a large multiplier rested on the idea that saving and consumption would be determined mainly by current income, but that seems not to be so: for example, it is argued that consumers will base their saving decisions not on this year's income but on their lifetime income prospects. But all of this refers to the short period of business cycles. When we address secular stagnation, however, we are looking at the longer run. Let us, for a moment, suspend our disbelief and consider the possibility that the relation between autonomous spending (averaged over a long period) and gross domestic product (averaged over a long period) might be determined by a large stable multiplier. Then, in given circumstances, the decrease in the rate of growth of investment would correspond to a decrease in the rate of growth of production and income, and a decline in investment to a decline in production and income. If, at the same time, the labor force and labor productivity continue to grow, this would imply a decline in employment and increasing unemployment – not as a temporary consequence of the business cycle but as a long-term trend.

Thus far we have assumed that stagnation occurs at a fixed aspirational profit rate. But aspirations are learned, not given forever. Thus, an adaptive economics (Chapter 6) would suppose that, as opportunities decline, aspirations would adapt by declining also. Rational expectations neoclassical economics would assume that they adapt instantly and optimally, so that the aspirational profit rate would be determined by a market equilibrium. In either case a lower aspirational profit rate would tend to offset the decline in investment, both by increasing the proportion of the current stream of opportunities that are fundable, and by making some of the backlog from earlier periods fundable. However, (1) taking informational costs into account, there will be a positive minimum rate below which profit-seeking entrepreneurs will not act, and the aspirational rate of return can go no lower. This minimum may have been reached some time ago. (2) To the extent that Keynesian positive feedbacks are predominant in disequilibrium states, adaptation may take the form of declining production and employment rather than decreasing aspirational profit rates. (3) To the extent that entrepreneurs rationally anticipate the consequences of continued stagnation, such as higher tax rates, this may lead them to shorten the payoff period, that is, raise, not lower, the aspirational profit rate. (4) There is little evidence that the aspirational profit rates of new entrepreneurs have declined in the late twentieth and early twenty-first century.

It may seem strange to make reference to secular stagnation in the twenty-first century. The term emerged in the 1930s to explain the persistence of low aggregate demand associated with low levels of investment. However, the term lost its currency in the third quarter of the twentieth century, a time of relative prosperity that, on its face, seemed to falsify the assertion of secular stagnation. However, (1) the postwar rebuilding of the economies of Europe and Japan created opportunities for investment that did not exist in the 1930s, a rebuilding process that continued through most of the third quarter of the century; (2) the economies of mature capitalist countries in that period were militarized, with government spending, deficits, and persistent inflation at levels that had no precedent in the earlier period, just as we would expect if governments were using fiscal and monetary policy to maintain aggregate demand against secular stagnation, (3) investment opportunities that have existed in the late twentieth century have often been consequences of government activities (such as the building of the Eisenhower Interstate Highway System and the expansion of post-secondary education in the USA), privatization of activities already carried on by government, or of epochal innovations, such as the cellular telephone and the internet, that have not arisen from competitive capitalist activity. The internet, we recall, originated from the US government and a European foundation, and the cellular telephone from what was at the time a regulated monopoly, the ATT Bell Laboratories. The cellular telephone might be taken as an illustrative example of what Schumpeter called "the routinization of innovation,"[(15)] one of the sources he saw of the decay of the special advantage of capitalism. Thus, we see in twenty-first century capitalism a system that relies on government to generate opportunities for entrepreneurs, a capitalism consistent with secular stagnation, rather than one in which such opportunities spontaneously exist to be found.

The relation of stagnation to privatization calls for a brief further comment. On the one hand, privatization can be a camouflage for deficit spending. If a private business sells a potentially income-earning property to cover its routine expenses, this would be recognized as a step toward bankruptcy. However, when a government sells an income-earning property, such as a water works, and spends the proceeds on routine expenses, this is no less an impoverishment of the organization of government. At the same time, it adds something to the stream of entrepreneurial opportunities – something that demands no creative response at all. It must be conceded that privatization has in some cases been the sale of losing properties, which were then capable of paying a profit as private enterprises. A typical case was the privatization of telephone services that had been required to cross-subsidize postal mail. When the privatization was accompanied by the elimination of cross-subsidies, this in itself should increase the profitability of the telephone service. In cases of this kind, we have a combination of deficit spending and tax cuts, along with the supplementation of the flow of entrepreneurial opportunities – which would offset secular stagnation even more strongly. Privatization, then, is a symptom of secular stagnation.

Another dimension of stagnation in the late twentieth century is the

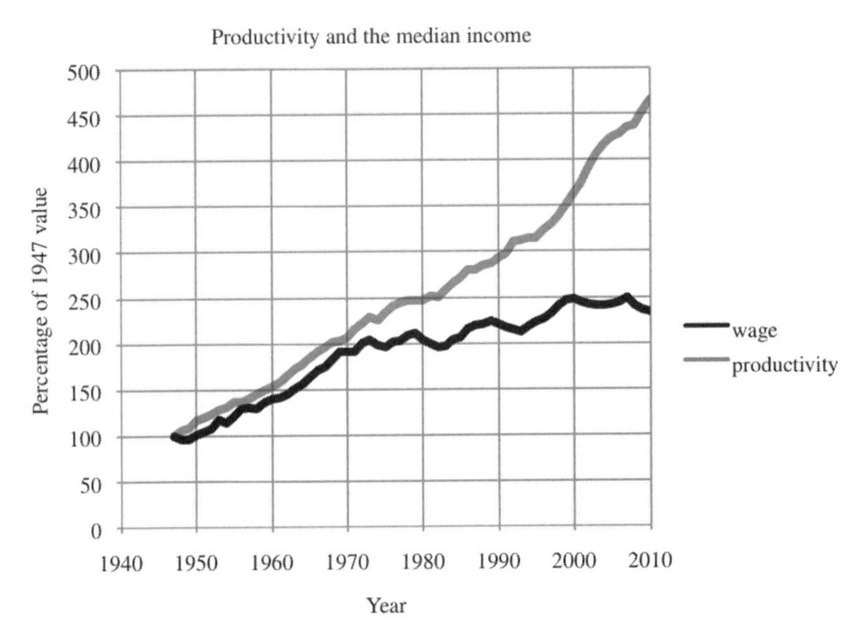

Figure 9.1 Trends of income and productivity

stagnation of real wages. It is widely argued that, at least in the United States, wages have stagnated since the 1970s. Some data and arguments that support this claim are illustrated by Figure 9.1.(16) In the figure, we see indices of labor productivity and median American family income, both adjusted for inflation. For a median family, income would mainly comprise wages at a representative wage rate. We see that the two series remain quite close before the mid-1970s, and diverge strongly thereafter, with median income remaining largely level while productivity continues to increase at a slightly slower rate.

In itself, secular stagnation might not explain the stagnation of wages. The immediate result of stagnation of entrepreneurial opportunities is stagnation of investment, and thus, through the Keynesian mechanism, of aggregate demand. This in turn would lead to a reduction in the competition of employers for labor, perhaps expressed by a rise in the "natural rate of unemployment," which would in turn restrain the rise of real wages. The details of this mechanism will have to be beyond the scope of this chapter, however.

Secular stagnation does, however, offer an explanation for another key dimension of current mature capitalist economies: the growth of government deficits and debts. In the face of secular stagnation of investment, fiscal deficits to maintain demand are a predictable government policy. Skeptical as many economists may be of simple Keynesian economic models – skeptical on what are essentially philosophic grounds of rational action theory – the gross evidence of recent history is that growing fiscal deficits are a consequence of governments struggling to maintain aggregate demand in a context of secular stagnation.

In this chapter we have explored the idea that a market economy may require guidance from the governmental mechanism in order to generate results that realize some approximation to the cooperative objective of the grand coalition of the whole population – that a market economy will unavoidably be a mixed economy. But is this mixed economy sustainable? It would appear that secular stagnation will require continually *increasing* fiscal deficits, and there seems to be some limit to the continuing increase in fiscal deficits. It may be, indeed, that the mixed market economy is not indefinitely sustainable. This is not to say that an *unguided* market economy would be sustainable either, supposing even that an unguided market economy is possible. Macroeconomic policy is a response to the failures of an unguided market economy to realize any reasonable approximation to the cooperative objectives of the population. If the guided market economy is unsustainable, then there would appear to be only two possibilities: the substitution of a mechanism based on non-market principles of some kind, or the failure of the grand coalition of the whole population.

CHAPTER SUMMARY

This chapter has continued the speculative discussion exploring the hypothesis that the population of a territory might form a grand coalition of the whole. In a world of costly information, this would undoubtedly require rules, hierarchies, and organization: in short, government. Nevertheless, the grand coalition might reasonably rely on non-cooperative decision-making in the framework of a market system, thus avoiding some of the costs of cooperative decision-making. This is an instance of imperfect cooperation, and as we saw in the previous chapter, the imperfections imply a large part of twentieth-century normative economics. In this chapter we have focused on macroeconomics, a body of policy aimed at stabilizing what may seem to be a rather unstable market mechanism. This, too, can be seen as a response to the imperfection of cooperation mediated by non-cooperative action in markets. However, the nature of that policy – which seems often to correspond to fiscal deficits, however opposed the dominant ideology may be – raises the question whether the grand coalition in this form can be indefinitely sustainable, and if not, what alternative there might be.

SOURCES AND READING

(1) Samuelson's original 1948 edition is of sufficient interest for the history of economics that it was republished in 1997: Paul A. Samuelson (1997), *Economics: The Original 1948 Edition*, New York: McGraw-Hill/Irwin. (2) Friedrich von Wieser (1891), "The Austrian School and the theory of value," *The Economic Journal*, **1**(1) (Mar.), 108–21. The quotation is at p. 108. (3) For Keynes see p. 161 of op. cit., Chapter 6 of this volume, note 18. For Malthus's writing on general gluts see Thomas R. Malthus (1836), *Principles of Political Economy*, London: W. Pickering, pp. 218–23. Although Keynes refers to Mill's version of the wages fund theory, Keynes seems to have been influenced by Mill's discussion in Essay 2 in John Stuart Mill (1874), op. cit., Chapter 1 of this volume, note 2, "On the influence of consumption on production," especially paragraph II. 71 et seq.

(4) The requirement of additional information to attain the strong equilibrium is the basis of a theory of indicative planning that I put forward in Roger A. McCain (1991) and Roger A. McCain (1985), both op. cit., Chapter 8 of this volume, note 19. I did not then understand the issue in terms of rationalizable strategies. However, the case is probably more complex in the actual world. See Chapter 5, especially sections 5.4 and 5.5 in Roger A. McCain (2009), op. cit., Chapter 1 of this volume, note 8.

(5) Early studies of the relation of consumption to income showed that it was more complex than Keynes supposed. James Duesenberry (1952), *Income, Savings and the Theory of Consumer Behavior*, Cambridge, MA: Harvard University Press, explained this mainly by habit formation and envy. Note also Robert H. Frank (2005), "The mysterious disappearance of James Duesenberry," *New York Times*, 9 June. Other scholars explained this in terms of far-sighted rationality. See Franco Modigliani and Richard E. Brumberg (1954), "Utility analysis of the consumption function," in Kenneth K. Kurihara (ed.), *Post-Keynesian Economics*, New Brunswick: Rutgers University Press, pp. 388–436 and Milton Friedman (1957), *A Theory of the Consumption Function*, Princeton, NJ: Princeton University Press. Each suggested that consumption would fluctuate less than current income, and the facts agree. However, there are limits to the ability of people to adapt their saving to their lifetime incomes, and as later research showed, consumption is not as stable as the models of Friedman and Modigliani predict. See especially Marjorie Flavin

(1981), "The adjustment of consumption to changing expectations about future income," *Journal of Political Economy*, **89**(5) (Oct.), 974–1009; (1985), "Excess sensitivity of consumption to current income: Liquidity constraints or myopia?" *Canadian Journal of Economics*, **18**(1) (Feb.), 117–36. (6) On plural Nash equilibria see Thomas Schelling (1960), op. cit., Chapter 1 of this volume, note 8. (7) As Keynes noted (e.g., op. cit. at p. 7), it was Arthur C. Pigou who did the most to make the neoclassical view of unemployment explicit, with its focus on labor markets: Arthur C. Pigou (1933), *Theory of Unemployment*, London: Routledge; reissued 1968. The 2010 Nobel Memorial Prize was awarded to Dale Mortensen, Christopher Pissarides, and Peter Diamond. See, for example, Dale T. Mortensen (1982), "The matching process as a non-cooperative bargaining game," in John J. McCall (ed.), *The Economics of Information and Uncertainty*, Chicago: University of Chicago Press; Dale T. Mortensen and Christopher Pissarides (1994), op. cit., Chapter 6 of this volume, note 3; and Peter A. Diamond (1982), "Aggregate demand management and search equilibrium," *Journal of Political Economy*, **90**(5), 81–94. As usual, the Nobel site is also a useful resource: http://www.nobelprize.org/nobel_prizes/economic-sciences/laureates/2010/, as of 17 November 2013.

(8) The simplest model that captures the idea of moving economic equilibrium is due to Nobel Laureate Robert Solow, from Robert Solow (1956), "A contribution to the theory of economic growth," *Quarterly Journal of Economics*, **70**(1) (Feb.), 65–95, and so is called the "Solow growth model." It begins with the Cobb-Douglas production function, originated by Professor, later Senator, Paul Douglas and Professor Charles Cobb, a mathematician. See Charles W. Cobb and Paul H. Douglas (1928), "A theory of production," *American Economic Review*, **18**(1), 139–65.

(9) The failure of bargaining theory in search-and-matching theory is documented by Robert Shimer (2005), "The cyclical behavior of equilibrium unemployment and vacancies," *American Economic Review*, **95**(1), 25–49. Nash's paper is John Nash (1953), "Two-person cooperative games," *Econometrica*, **21**(1) (Jan.), 128–40. On bargaining and bargaining power see Roger A. McCain, op. cit., Chapter 3 of this volume, note 15; on Nash's non-cooperative bargaining model see in particular Chapter 3 of this volume. The contrast between "non-cooperative" and other bargaining models is discussed at an introductory level in Chapter 17 of Roger A. McCain (2014), op. cit., Chapter 1 of this volume, note 8. For evidence on unequal bargaining power see, for example, Jan Svejnar (1986), "Bargaining power, fear of disagreement, and wage settlements: Theory and evidence from US industry," *Econometrica*, **54**(5) (Sep.), 1055–78; Matthew Grennan (2013), "Price discrimination and bargaining:

Empirical evidence from medical devices," *American Economic Review*, **103**(1), 145–77.

(10) Efficiency wage theory emerged in the study of less developed countries and focused on the idea that higher wages could lead to improved nutrition and so to higher productivity. See Benjamin Higgins (1968), *Economic Development*, revised edition, New York: Norton, pp. 320–24; Joseph Stiglitz (1976), "The efficiency wage hypothesis, surplus labour, and the distribution of income in L.D.C.s," *Oxford Economic Papers*, **28**(2) (Dec.), 185–207. For an early general statement of an efficiency wage theory for industrialized economies see James M. Malcomson (1981), "Unemployment and the efficiency wage hypothesis," *The Economic Journal*, **91**(364), 848–66. There are several lines of thought from which an efficiency wage model can be derived. George A. Akerlof (1982), "Labor contracts as a partial gift exchange," *Quarterly Journal of Economics*, **98**(4) (Nov.), 543–70 focuses on reciprocity. For the interpretation in terms of repeated games, see W. Bentley MacLeod and James M. Malcomson (1993), "Wage premiums and profit maximization in efficiency wage models," *European Economic Review*, **37**(6) (Aug.), 1223–49. Christopher Martin (1997), "Efficiency wages: combining the shirking and turnover cost models," *Economics Letters*, **57**(3) (Dec.), 327–30 synthesizes two other approaches.

(11) On customers in a search and matching model see Robert E. Hall (2008), "General equilibrium with customer relationships: A dynamic analysis of rent-seeking," Hoover Institution and Department of Economics, Stanford University, http://www.stanford.edu/~rehall/GECR_no_derivs.pdf, as of 17 November 2013. It seems that under conditions of wartime rationing, merchants sometimes favored long-term customers to the extent of excluding new ones, and this was seen as a problem: W. Arthur Lewis (1942), "Notes on the economics of loyalty," *Economica*, New Series, **9**(36) (Nov.), 333–48.

(12) The charge of reification is made, for example, by James Kenneth Galbraith (2001), "How the economists got it wrong," *The American Prospect* (Dec.), available at http://prospect.org/article/how-economists-got-it-wrong, as of 6 July 2013. (13) See Joseph A. Schumpeter (1950), *Capitalism, Socialism and Democracy*, 3rd edition, New York: Harper. (14) The secular stagnation hypothesis seems to originate with Alvin H. Hansen (1939), "Economic progress and declining population growth," *The American Economic Review*, **29**(1) (Mar.), 1–15. For a more recent discussion see John Bellamy Foster (1997), "The long stagnation and the class struggle," *Journal of Economic Issues*, **31**(2) (June), 445–51. While this book was in press, Larry Summers was reported to have suggested that the United States might be in a state of secular stagnation in the present and

foreseeable future. According to a report in Bloomberg Business Week, his remarks were made "at a Nov. 8 speech at the International Monetary Fund;" however, Summers did not confirm this. See McCoy, Peter and Matthew Phillips (2013), "Mr. Negative," Bloomberg Businessweek, (Nov. 25) pp. 13-14. The references to Schumpeter are to Joseph A. Schumpeter (1947), op. cit., Chapter 6 of this volume, note 20; Joseph A. Schumpeter, Markus C. Becker and Thorbjørn Knudsen (2005), op. cit., Chapter 8 of this volume, note 13. (15) On epochal innovations see Simon Kuznets (1960), *Modern Economic Growth, Rate, Structure, and Spread*, New Haven, CT: Yale University Press. On routinization of innovations, see pp. 132–4 of Joseph A. Schumpeter (1950), op. cit., note 13 above.

(16) Diagram prepared by the author, using data from Economic Policy Institute (2013), "The state of working America," available at http://stateofworkingamerica.org, as of 31 July 2013, and from the Bureau of Labor Statistics. See also Susan Fleck, John Glaser and Shawn Sprague (2011), "The compensation–productivity gap: A visual essay," Bureau of Labor Statistics, accessed 21 March 2013 at http://www.bls.gov/opub/mlr/2011/01/art3full.pdf. Steven Greenhouse (2013), "Our economic pickle," *New York Times*, 14 Sunday January, p. SR5, summarizes some of the data and arguments supporting this view, as well as some qualifications.

10. Political economy

Chapter 8 speculated about a grand coalition of the whole society, and how it might adapt to the cost of information. We found that this hypothetical grand coalition would not be able to rely exclusively on incentive-compatible rules, but would operate as an organization, with an agent or agency of some sort to enforce an approximation to a cooperative solution for the whole society. Chapter 9 continued the speculation with a focus on what are known as macroeconomic policies. If we suggest that actual governments act as such agents, then we are paralleling the ideas of the social contract theorists in political philosophy. There are some differences. One major point of social contract theory is the normative argument that we are under a moral obligation to obey the state. The purpose of this work is not to draw conclusions about moral obligation but to understand the mixture of cooperative and non-cooperative action that we observe in the actual world. Nevertheless, the insights of the social contract theorists will be useful to us, and particularly those of the founder of the western social contract tradition, Thomas Hobbes.

A HOBBESIAN GAME

Hobbes argued that rational individuals would obey a monarch *in their own interest.* Hobbes[(1)] writes (1651, p. 80):

> From this fundamental law of nature, by which men are commanded to endeavour peace, is derived this second law: that a man be willing, when others are so too, as far forth as for peace and defence of himself he shall think it necessary, to lay down this right to all things; and be contented with so much liberty against other men as he would allow other men against himself. For as long as every man holdeth this right, of doing anything he liketh; so long are all men in the condition of war.

The Hobbesian Game, shown in Table 10.1, is based on these ideas although it is, of course, very much oversimplified. There are just two players, William and James. (This is a particularly important oversimplification, but nevertheless, let us see what the game can teach us.) The payoffs represent the satisfaction each person has from life, on a scale of 0–5.

Each player has three possible strategies: obey King Henry, obey Prince John, or obey nobody. The lower right cell, where each obeys nobody, is Hobbes's state of nature: "that condition which is called war; and such a war as is of every man against every man . . . and the life of man, solitary, poor, nasty, brutish, and short" (ibid., pp. 77–8). Accordingly the payoffs are the worst, at zero. Where both obey the same overlord, they live in peace, and this is the best for both of them. Where they obey different lords, they are in a condition of civil war – not quite as bad as the state of nature, but quite bad all the same. Finally, where just one obeys nobody, his liberty will be repressed by the organized state (which consists of the other player and the overlord) but the repression is costly. Thus, the player who obeys nobody is again in the worst position, while the player who is part of the organized state does a little better.

Table 10.1 A Hobbesian Game

First Payoff to William, Second Payoff to James		James		
		Obey King Henry	Obey Prince John	Obey nobody
William	Obey King Henry	5,5	1,1	2,0
	Obey Prince John	1,1	5,5	2,0
	Obey nobody	0,2	0,2	0,0

This game has two non-cooperative equilibria. The Nash equilibria exist where the two players each obey the same overlord. (In other words, this is a coordination game.) In this sense, it is in the interest of each individual to be part of the organized state, under one overlord or another. But the cooperative solution for this game is also that both obey the same overlord. In short, the Hobbesian Game is one of those special cases in which a cooperative solution is also a non-cooperative solution. This is the good news, but there is some bad news. Both of the strategies of obeying one overlord or another are rationalizable, so rationalization does not get us very far and may leave us in a very bad position. We need a bit more information to assure that one of these non-cooperative/cooperative equilibria will be realized. This information might come from social customs, such as the convention that the succession to the monarchy is hereditary. Historically, such a customary norm has been quite common, but has also failed from time to time, resulting in civil wars.

As the reader was warned, the Hobbesian Game is very simplified by comparison with Hobbes's complex and nuanced thought. In reality (and in Hobbes), the number of participants is not two but is very large. Nevertheless, what is true in the simplified two-person game also seems to be true in the world: for the vast majority of people, it is in the interest of each to be a part of an organized state, and this self-interested commitment by many individuals creates "that great Leviathan, [who] by this authority, given him by every particular man in the Commonwealth, he hath the use of so much power and strength" (ibid., p. 106) that is able "keep them all in awe" (ibid., p. 104). That is, the monarch's power is usually sufficient to prevent individual or cooperative deviations from the grand coalition.

But we need to point out another aspect of Hobbes's idea: the overlord is not a member of the cooperative coalition. He or she is, in a sense, the agent of the coalition but not a member of it. The non-cooperative equilibrium at which the residents of the country all choose to obey the overlord imposes no obligations on the overlord. In particular the monarch is under no obligation to respect the rights of the citizens. This is Hobbes's argument that the social contract implies an absolute government. Put otherwise, the non-cooperative agreement that Hobbes envisions is one that leaves all bargaining power in the possession of the overlord, and none to citizens of any class. Later social contract theorists questioned that, and once again, we can gain some insights from exploring their ideas in a game-theoretic way.

A LOCKE-ROUSSEAU GAME

Later social contract theorists reconsidered Hobbes's absolutism. Locke(1) argued that the social contract might take the form of a system of laws or practices designed to protect the life and property and civil liberty of the citizens. We shall call this a system of liberal laws. But liberal laws are not the only possibility. The power of the organized state may be corrupted and corrupting, and the laws and customs that comprise an existing social contract might actually leave people worse off than they would be in the state of nature. This idea was stressed by Rousseau.(1) Let us call a legal system of this kind repressive laws. Thus, our two decision-makers have to choose among three strategies: one strategy is to consent to liberal laws, a second to consent to repressive laws, and finally, again, to reject all laws. If both reject all laws, we again have the state of nature. Locke is less pessimistic about the state of nature than Hobbes is, but argues that in the conditions of his time, a state of nature without a government power would be less convenient than an organized state. These ideas are reflected in our oversimplified way in the payoffs at the upper left and lower right of the Locke-Rousseau Game, Table 10.2. The off-diagonal payoffs are unchanged from the Hobbesian Game and follow from similar reasoning. We can see that there is a Nash equilibrium at which both decision-makers choose to consent to liberal laws, and this also is the cooperative solution in this very simple game. Thus, we can say again that in this game, the cooperative solution is also at the same time a non-cooperative equilibrium.

Table 10.2 A Locke-Rousseau Game

First Payoff to William, Second Payoff to James		James		
		Consent to liberal laws	Consent to repressive laws	Reject laws
William	Consent to liberal laws	5,5	1,1	2,0
	Consent to repressive laws	1,1	3,3	2,0
	Reject laws	0,2	0,2	4,4

It might seem strange that people would choose the strategy of consenting to a system of repressive laws, but we see that they will be better off than they would be in a condition of civil war, when the two consent to different legal systems, or than they would be if brought into the organized state by force, against their decision to reject all laws. Thus, repressive laws can also be a Nash equilibrium, as the mild state of nature also is. We see that the Locke-Rousseau Game has three Nash equilibria. Each case in which both decision-makers choose the same strategy is a Nash equilibrium. We may apply Robert Aumann's refinement of Nash equilibrium, and look for a strong equilibrium,(1) that is, the Nash equilibrium that the two decision-makers would agree on if they sit down and talk about it. This would be the Nash equilibrium at the upper left. However, all of the strategies in this game are rationalizable. Thus, rationalization might lead to any combination of strategies, and in particular, if the repressive laws are customary, each person might rationalize that the other one would choose the customary strategy of repressive laws. To be more exact, William might reflect: "James thinks I will consent to repressive laws, since it has been his experience that others do so, and because he believes that, he will consent to the repressive laws – so my best response is to do the same." This will be particularly likely in the more realistic game with many participants, where repressive laws have long existed. This would result in the Nash equilibrium at the repressive laws.

Something more needs to be said about the simplifying assumptions in this game example. Once again, we have used an example of a two-person game to represent what in actual fact is an interaction of millions of decision-makers. While the many-person game may have similar Nash equilibria, the large numbers in the actual world will find it all the more difficult to attain the strong equilibrium at the upper left, since this large group of people cannot sit down together and talk it out. But a second simplifying assumption is that the two players in each of these two games are identical, and obtain the same payoffs in every Nash equilibrium. In reality there will be individual differences among people and consequently different types of players. In particular, individual payoffs for the different strategies may differ from the averages shown in the table. This may change their preferences among the Nash equilibria. Some, who are strong, clever, and ruthless, may be better off in the state of nature than in any organized state (although Hobbes argues against this possibility; op. cit., p. 76). In the case of repressive laws, a minority will benefit from them, and be better off with repressive than with liberal laws. This is an intuition toward a theory of revolution. In the game, if both William and James withdraw their consent to the repressive laws, then the repressive laws no longer have any force, and the liberal laws – or a state of nature – may be painlessly substituted for them. In reality, of course, it is not so simple. The minority who benefit from repressive laws will have the ability to resist such a passive revolution, so that, when a large proportion of society withdraw their consent to the existing legal system, the result is more likely to be civil war. There do seem to be a few exceptions, where non-violent non-cooperation has brought about a shift from more to less repressive laws and customs without civil war, but this is far from the simplicity of the two-person game example.

What we learn from these examples, loosely based on the ideas of the social contract theorists, is that a cooperative solution to the game of organizing a political system can also be a non-cooperative solution. Once we arrive at the cooperative solution it is self-enforcing, as any Nash equilibrium is self-enforcing. But it may not be the only non-cooperative solution. In a many-person Hobbesian or Locke-Rousseau Game, as we have seen, individual differences may put different people in different situations with respect to the legal systems that would favor them. When we recognize this we have, in effect, recognized differences of class. This is another point on which the simplicity of these games, and the social contract theories, may be criticized. Perhaps the social contract is not a coalition of the whole society at all, but rather a coalition of one class at the expense of another. And this, of course, is one interpretation of the ideas of Marx and Engels.

CAPITALISM AND DEMOCRACY

In the previous chapter we speculated on how a hypothetical grand coalition of the whole society might be organized, seeking an approximately cooperative solution while limiting the cost of information. One possibility was found that resembles the democratic market economies that are common many industrialized modern nations. Can this be more than an incidental resemblance? The political economies of modern nations are products of their history. Can we find anything in their histories that might account for this resemblance? One thing that is suggestive is that all are mature capitalist countries. The maturation of capitalism has corresponded to the emergence and spread of representative democracy. How are these trends connected?

In matters such as these, definitions of terms may themselves be objects of contention. For Marx and Engels,(2) who originated the term, capitalism was understood to be a class system, the most recent form of class society, in which key roles are played by an owning and employing class and a class of persons who rely on the sale of their labor to the employing class, that is, a working class. Schumpeter agrees. To be specific, however, "Unlike the class of Feudal Lords, the commercial and industrial bourgeoisie rose by business success . . . in most cases the man who rises first into the business class and then within it is an able businessman."(3) Thus, we will define capitalism as: a class system in which key dynamic roles are played by a class of owner-employers and a class of employees, and in which social advantage, including membership in the employing class and social position within it, are consequences of business success.

Something needs to be said about the relation of capitalism, in this sense, to a market economy. To say that social position is a consequence of business success requires that business success is itself judged on some basis that is neutral to social position. A market system provides such a neutral judgment of business success. This does not require that the market system be characterized by laissez faire or perfect competition. Broadly based taxes, such as tariffs or green taxes might distort the allocation of resources but would not seem to bias the judgment of business success. Even corruption might not, if its incidence were random or like a broad-based tax with a random component. Thus, a market system that evaluates business success is a necessary basis for a capitalist class, but the interest of any individual capitalist would be advanced by a system that would favor business people of her or his own group. In this sense, neutral markets are in the class interest of capitalists but not in the interest of any individual capitalist.

The term "working class" is also contentious. For the purposes of this chapter, as in Marxist thinking, the working class comprises all those who rely on the sale of labor power for an income, over most of the life cycle, *whether they consider themselves members of the working class or not.* This can be confusing in a country like Britain, where there is a distinct "working class culture," or like the United States, where virtually nobody thinks of him- or herself as a member of the working class; but for the purposes of this chapter a clear definition is necessary and this is the one that fits best with the purposes of the chapter.

For the purposes of this chapter, the meanings of "democracy" and "market system" are the ones discussed in Chapter 8. Why, then, do we see the historical association between capitalism and democracy in the senses associated to them? In the Communist Manifesto, Marx and Engels write, "The executive of the modern state is but a committee for managing the common affairs of the whole bourgeoisie."(2) Even to the extent that it was true in 1848, this is an oversimplification that enables Marx and Engels to avoid a difficult problem with respect to capitalist societies as political systems. In a feudal system, it is the military force of the landlords and their armed retainers who enforce the rights of the aristocracy, suppress banditry and impose order on their domains. In a classical city, an urban tributary society,(4) it is the citizens in arms, as a body, who enforce their tributary supremacy over lesser cities and the countryside. The aristocrats of a feudal system and the citizens of an urban tributary system are governing classes in a literal sense. Capitalists are not. In a capitalist country, soldiers and police officers are public employees. Within capitalism, social advancement comes from business success. If the members of the "committee for managing the common affairs of the whole bourgeoisie" are themselves capitalists, then either they are neglecting their own businesses or they are using their political position to increase their profits. In the former case they are likely soon to cease to be capitalists. In the latter case, crony capitalism, it is no longer the case that social advancement comes from business success, but on the contrary, business success comes from social advancement in the form of government connections.(5)

Nor can strong government be considered an imperfection somehow alien to capitalism. Historically, the emergence of capitalism coincided with the establishment of some of the strongest, most centralized governments in human history. Can this be a coincidence? It is not: a society in which social position comes from business success demands *uniform* enforcement of property rights and contracts, and suppression of banditry, fraud, and coercion over a wide territory. It is a mistake to underestimate the government power necessary to achieve that. Moreover, these requirements for the viability of a capitalist system and a capitalist class often conflict with the interests of individual capitalists. As we have seen, capitalism creates a series of Prisoner's Dilemmas that can only be resolved through government coercion. A government "small enough to drown in a basin" would be much too small to sustain capitalism.

Who, then, shall govern in the absence of a governing class? One possibility is a strong man, a Cromwell or Bonaparte or Pinochet, and as the examples will suggest, this has often been tried. From the point of view of the capitalist class, this has two shortcomings. One is that strong-man government alternates with periods of civil violence, often at the time of succession. The other is the converse: the strong man will often consolidate his rule at home by means of war abroad. And this is destructive of the interests of the capitalist class. Nothing is more certainly in the interest of a capitalist class than peace at home and abroad.(6) This observation will be a key assumption in what follows. Even in peacetime, strong-man government seems often to disrupt market processes, which are, as noted above, a necessary basis of capitalist society.

On the other hand, an electoral system, with a property test for the franchise sufficiently strict that only capitalists may vote, provides a "committee for managing the common affairs of the whole bourgeoisie" without succession struggles nor any predictable tendency to make war to stabilize or extend the power of the autocrat. Government is then left to professionals in the field, who must appeal to a majority of the capitalist class by governing in their interest. Such a system has been more successful in maintaining peace at home and abroad than strong-man rule has, while at the same time it subordinates the government to the interests of the capitalist class as a whole. And as Marx and Engels were writing, in 1848, it was the most likely form that an electoral system would take. But as Bernstein notes, the property test was not itself stable.(7)

In 1848 and the period following, Marx and Engels understood that the working class was then a minority class, as Engels (1895) later wrote.(2) In 1848, in post-Bonaparte Europe, the liberated peasants probably were "the poorest and most numerous class," and there were remnants of the medieval artisanat. In this context Marx and Engels predicted that the working class would grow to be a large majority, subsuming those other groups; that its political power would increase, and that a revolution of the working class would make it the dominant class and result in the abolition of the capitalist class. The last of these three expectations has not been realized.* But the first two were, and such contemporary ideas

* Revolutions in countries like Russia and China are not exceptions to this statement, since the prediction of revolution in Marx and Engels's ideas was a prediction about the mature capitalist systems. The twentieth-century revolutions in countries where immature capitalism was mixed with feudal and pre-feudal elements, and the rise and decline of colonial empires, important as they are, are beyond the scope of this book.

as the classical stationary state had no such qualified success. As Engels (1895) describes it,[2] and as Bernstein emphasizes,[7] the power of the working class *within capitalism* increased with the continued expansion of capitalism throughout the latter half of the nineteenth century. The forms that supported that shift of power toward the working class – workers' party participation in parliaments, organization of labor unions, and consumer cooperatives – were not forms that Marx and Engels anticipated in 1848, and were forms that Marxists often opposed as diversions from the revolutionary project. But at the same time the predicted expansion of the working class and the diminishment of the other classes created the possibility that the working class revolution, for the first time in history, could be a majority revolution. That, at least, was Engels's perspective in 1895.

CLASS COMPROMISE

What had emerged by the beginning of the twentieth century was something Marx and Engels certainly had not anticipated: a class compromise between the two principal classes in the capitalist systems that were then relatively mature. The compromise had different terms in different countries, and evolved over the first half of the twentieth century. By 1900, a social security system was already an important part of it in Germany; in other countries, manhood suffrage and some tolerance of labor unions played important parts. Later, in some English-speaking and other countries, socialized medicine became an important part of the compromise. More or less progressive tax systems are also a part of the class compromise wherever we find it. In every case where the compromise took root, representative democracy was a part of it.[8]

Marx and Engels could not have anticipated such a compromise, because they supposed that the interests of the capitalist and working classes were completely opposed. But this was probably an error. In a complex world, the interests of the classes may be fundamentally opposed, but there may be many details in which enough common interest can be found to support a compromise. Public policies to limit cyclical unemployment provide one instance. The German social security system seems to be another. It was cheap, since at that time few workers lived long enough to collect social security, and indeed, insofar as the state acted as an insurer to the working class, pooling the individual risks of poverty in old age along lines suggested by Chapter 5, it probably cost the employing class nothing, while benefiting the working class.

Beyond these matters of detail, though, it is likely that Marx and Engels were wrong (in the context of the nineteenth century) in seeing the interests of the employing and working classes as fundamentally opposed. The limited literature on class compromise in political sociology begins from a game theoretic model due to economist Kelvin Lancaster.[(8)] In Lancaster's highly simplified model, a game is played over time by two classes. One of these classes, the employing class, determines what proportion of profits will be invested. The other player, the working class, determines what proportion of production will go to profits, within an upper limit determined by competition among the capitalists. Lancaster reasons that the working class, because of its numbers, has the power to expropriate the capitalists at will, reducing their profits to zero. However, it is initially not in the interest of the working class to do so, since the reinvestment of profits increases labor productivity and so raises wages. However, Lancaster incorporates diminishing returns in an extreme form: labor productivity can be increased linearly only up to some maximum. Once its maximum is reached, it is no longer in the interest of the working class for the capitalist process to continue, and expropriation takes place. But in fact this maximum is never reached. Anticipating that their wealth will be expropriated, capitalists find it in their interest to stop reinvesting profits before labor productivity reaches its maximum, and at that point the rise in wages stops, so that it is no longer in the interest of the working class for the capitalist process to continue. Thus, the cessation of reinvestment of profits, a result of capitalists' anticipation of a future crises, results in the crisis they anticipate: expropriation. Thus, the Nash equilibrium of this game, Lancaster says, is dynamically inefficient.

Clearly this is a highly stylized model. We might think that the role of capitalists is not only to reinvest but also to search intensely for profitable, innovative opportunities in which to invest, or, in the terms of earlier chapters, to produce information that may lead to the emergence of new technologies and organizations. Conversely, to the extent that diminishing returns exist, they may be more complex and gradual than Lancaster assumes. This would make the game more difficult to solve mathematically. Moreover, classes are not unitary, rational players of a two-person game. At best they are rather dumb beasts with only confused notions of their interests; at worst, perhaps a class has more in common with a cancerous tumor than with a poker player. Lancaster probably exaggerates the farsight of the capitalist class, since in practice the pursuit of class interest is an evolutionary process and we might expect the shorter-term interests of classes to be the effective ones. Even so, as Schumpeter says,[(3)] "historical events may often be interpreted in terms of class interests," and Lancaster's model is useful for what it tells us about class interests. In particular, at a certain stage in the evolution of capitalism, it may be in the interest of the working class that the capitalist process continues; while at a later stage this may no longer be true.

One further point of common interest between the two major classes is peace. In war, foreign or civil, wealth is destroyed and blood is shed, and in the nature of a capitalist society, the wealth that is destroyed is the property of capitalists, for the most part, and the blood that is shed is shed by members of the working class, for the most part.

These ideas, like our other game theory examples in this chapter, are highly simplified, but they suggest that in 1900 or before there was ample basis for a compromise between the working class and the bourgeoisie. The mature capitalist economies of the twentieth century had two characteristics that seem to distinguish them from many, if not all, earlier class societies. First, the most powerful classes within those societies are distinguished by their economic function, not by their possession and deployment of the means of violence. Second, class power in these societies was divided, significantly if not symmetrically, between two such classes. Both of these characteristics favor compromise and the formation of a grand coalition of both classes, and so, in effect, of the entire society. Moreover, such a grand coalition is one of the Nash equilibria of the underlying Hobbes-Locke-Rousseau non-cooperative game of state formation. Thus, once attained, this grand coalition can be stable in a non-cooperative sense: both a non-cooperative and a cooperative system. But it is not the only Nash equilibrium of the Hobbes-Locke-Rousseau game. During the twentieth century some mature capitalist countries adopted class compromise and some rejected it. The rejection of class compromise was associated with fascism. Those that adopted class compromise were, on the whole, successful as capitalist countries. In the others, there has been less success and many such regimes have been terminated, often in war or social conflict. Whether this will be the case also in the twenty-first century can only be an open question.

The balance of class power is a real and important reason for the emergence of market economies with representative government, but the picture we have drawn in this chapter is still a simplified one. Classes are not monoliths, and a government based on class compromise is open to other kinds of compromise as well.

PLURALISTIC COMPROMISE

On the face of it, it would *not* seem that a system of majority rule would operate in the interests of the capitalist class. After all, a capitalist class can only be a minority class. This is a matter of mathematics really: if wealth ownership is distributed among the individuals in society, the group who possess the largest holdings of wealth will be relatively few in number. Conversely, the working class includes all those who rely on the sale of their labor power, and they are a large majority in a mature capitalist system. Yet the class interests of the working class do not always decide public policy. This leads some theorists to the conclusion that class compromise can only be a sham.

The growth of the working class to be a large majority did not deliver plenary power into the hands of the working class, as simply counting heads might suggest. As William Riker (1962)[9] pointed out, drawing on cooperative game theory, large majorities tend to automatically splinter in a system of majority rule. The ideal size of a winning coalition is just 50 percent +1. Larger coalitions have to divide the limited gains from their majority position among more heads, so the smallest winning coalition is likely to be the one that most benefits its members.

A working person is not just that: she also is a person of a particular ethnic group and religious convictions, living in a particular region, with her own tastes and preferences and pastimes. This can equally be said of a member of the capitalist class, but its significance for the working class arises from the fact that the working class is a large majority. Precisely because it is a large majority, the working class is divided into factions in a majoritarian system. Thus, for example, religious parties in Europe appeal largely to working people for whom religion is relatively important.

Mancur Olson's (1971)[9] ideas take this even further. Olson thinks in terms of non-cooperative rationality and, while he may have been influenced by game theory, his ideas are not formally set out in game-theoretic terms. However, the cost of information is a key to them. Olson argues that small, closely connected minorities can have an influence on political outcomes that is very much out of proportion to their numbers, and often can have decisive influence. This is a result of the cost of information, in that the information cost of forming coalitions rises with the number of people in the coalition. As we have seen, inefficient externalities are explained partly on that basis, and we would expect the cost of information to be no less important in determining what political coalitions are formed and are effective.

These tendencies undermine the decisive power of a large majority and are often thought of as imperfections in democracy. It certainly is true that the politics of ethnic division has resulted in some of the most vicious and tragic political events of the twentieth century. However, these have largely occurred in countries that rejected class compromise for absolutism. Even within the context of representative democracy, the ability of compact minorities to organize as lobbies has promoted some regrettable public policies. The subsidies to farmers, noted in Chapter 8, would be one example. And, from the point of view of activists for the working class, this limitation of the bargaining power of a large majority is at best a problem to be solved.

However, there is some good news. Within the context of a class compromise, with the liberal laws and protection of civil rights that class compromise seems to promote, the electoral system is transformed from one of compromise between classes to one of compromise among various factions, including the capitalist class as one faction. This may well be a condition necessary to enlist the capitalist class in the class compromise itself. The resulting openness to a variety of interests and concerns apart from class interests can also be a good thing in itself. This can be illustrated by the example of environmental politics.

Environmental issues in politics arise from negative externalities[(10)] in the form of pollution or exhaustion of environmental resources. These are inefficient externalities in the sense that, while some groups gain from the activities that generate the externalities, others lose from them, and the losers lose more than the gainers gain. In a market system, since we rely on non-cooperative decision-making in non-government economic decisions, the groups who gain will pursue their own profits and generate the externalities. This will also influence the production of information, so that new technologies developed will often be the ones that increase, rather than decrease, the externality. Government policies to limit or eliminate these externalities are policies that internalize the externality, that is, that require those who generate negative externalities to make a money payment equivalent to the losses on the part of those who are victims of the externality. The best-known such policy was suggested by Pigou, who originated the concept of externality. His proposal is that a tax be imposed on the activities that generate negative externalities, and the tax be just large enough to internalize the externality. Such a Pigovian tax is recognized by economists of many different orientations as a policy to bring about efficient use of resources despite the presence of (potentially) inefficient externalities.

Is this a matter of class interest? The answer is no, for two reasons. First, internalization of externalities is an increase in efficiency, so that it generates more benefits than costs. In a world of free information, the benefits from the efficient policies would be used to compensate the net losers with lump sum money payments. Neither class has anything to lose.

Second, in the actual world of costly information, lump sum payments are not a possibility. The transition from an inefficient situation to an efficient limitation of activities that generate environmental externalities will have some informational costs, as some investors who have profited from those activities and some workers who have been employed in them shift to other, non-polluting activities. These informational costs in the first instance fall on *those* investors and workers, not on others. Thus, some capitalists, and some workers, have something to lose from that transition. In both classes, there are both beneficiaries and losers from the transition.

Some would say that, nevertheless, environmental issues are working class issues, on one or both of two arguments. The first is that the losers from environmental externalities are disproportionately the poor. However, as we have just seen, impacts on working people will vary by industry and location; a disproportionate influence will not necessarily create a common interest among the members of the class. The second argument is that since the capitalists take the profit, workers cannot benefit from externalities and thus cannot lose by policies that rectify them. This argument corresponds to classical and neoclassical economics, in which by assumption capitalists maximize profits and employees get only their second-best alternative, but misses two points. First, in a real capitalist enterprise, employees will usually have some bargaining power and so get some share of the surplus generated by the enterprise. Second, this classical/neoclassical argument ignores the informational cost of shifting to a new job in a non-polluting industry.

Mature capitalist countries have made some, though limited, progress toward efficient resolution of environmental problems. Probably the divisions within both of the most important classes account for the relatively slow progress on this issue. On the other hand, without the civil liberties and electoral government that have their own basis in class compromise, the politics of environmental issues would surely take a different form, and might not be possible at all.

CAN CLASS COMPROMISE BE SUSTAINED?

The most successful of mature capitalist countries in the twentieth century have based their success on class compromise, as we have seen. But may we expect that this will continue to be the case in the twenty-first, and if not, can some other direction be foreseen?

Sustainability

Lancaster's model of class compromise,(8) we recall, assumed that class compromise would be in the interest of the working class so long as wages would continue to rise. But, as we saw in the previous chapter, there is evidence that (in the United States, at least) wages have lagged behind productivity and have more or less stagnated. This may be a consequence of secular stagnation, that is, a slowing of the capitalist entrepreneurial process as it can no longer harvest the low-hanging fruit. Whatever the reason for the stagnation of wages, it removes one major advantage for the working class in the class compromise, and arguably the most important one.

While they are differently provided, the various components of the social safety net are also an important part of the class compromise of mature capitalism. Assurance of health care, for example, may be more generally provided by the government in some cases or by the employer by others, but in either case it generally is in the interest of the working class and so an element of class compromise. However, two key elements of this compromise have absorbed increasing proportions of the national products of mature capitalist countries. They are health care and pensions.

The reasons for the growth of health care expenditures are widely debated. There is no question that the health care system in the United States is remarkably inefficient, but there is no evidence that this inefficiency has increased in recent decades, and there is evidence that other countries, with quite different health care systems, have also experienced increasing costs. William Baumol and his associates(11) have offered an explanation in terms of the characteristics of demand and technical progress in medical care, and this explanation seems to fit the facts. Baumol's theory tells us that expenditure for the *efficient* provision of health care will require an increasing proportion of total product as time goes on, so that (1) elimination of inefficiencies can only, at best, slow this increase, not stop it, and (2) to the extent that they are approximately efficient, all systems for health care provision, public or private, will see increasing costs. Baumol argues that the same economic facts account for the rising cost of education, which also plays a key part in the class compromise of the twentieth century, though national systems vary widely.

The reasons for increasing pension costs are less controversial. They can be found in demographic change – the decreasing growth of the population (which in some mature capitalist countries is negative) and the consequent aging of the population. Here, again, national systems differ – the United States relies more on private pensions than European countries do, and has shifted the tax burden of its social security system backward, shifting from an intergenerational transfer system to a trust fund system, to a considerable degree. European countries have not. In any case, the charge of pensions against national product will continue to increase, however the systems are funded.

All of this is complicated by the growth of government indebtedness and deficits, and this is especially troublesome in the Euro area, where the member countries cannot monetize their public debt. There is no *direct* connection between rising public expenditure and public debt. Taxes can be raised to pay for public expenditures whatever they may be. If information were free, these might be lump sum taxes, which would have no impact on efficiency. In practice all taxes do affect efficiency, and usually negatively. In the terms of neoclassical economics, these taxes have an excess burden, and the excess burden of taxes to pay for beneficial public programs should be charged against their benefits in cost–benefit analyses.

Some editorial writers and others argue, however, that the deficit is a *result* of the growth of spending. The assumptions behind this view are often not made very explicit, but the assumption seems to be that representative governments (or all governments) tend to increase their expenditure without regard to efficiency. The idea seems to be that voters treat the establishment of the programs and the creation of taxes to pay for them as independent decisions, and hope to benefit from the programs without bearing the cost in taxes. Conversely (this view assumes) the voters suppose that they can benefit from tax cuts without losing the programs that they like. As voters base their votes on these suppositions, the representatives gain by basing their policies on them. But, of course, the programs must be paid for, so that increasing government expenditure is against the interests of almost everybody, according to this view. These writers tend to borrow Hobbes's term "Leviathan" to refer to this malignant growth of government, which is a bit odd in that Hobbes argued that the aggrandizement of government is in, *not* against, the interest of almost everybody. In any case, this view sees government deficits and "excessive government spending" as the symptoms of systematic inefficiencies of representative government.

There is no evidence that voters are unaware of the costs of the government programs they support. The Ricardian equivalence principle[(12)] tells us that people are rational and foresightful enough to anticipate that current government deficits will be paid for by future tax increases, and to hedge against the increase. If they are only boundedly rational and not rich enough to invest in hedging, then they may do this quite imperfectly. But whatever they can foresee when they invest they can foresee when they vote. This chapter has argued that the class compromise of representative government will make decisions that are approximately efficient, and even though the approximation is only rough, the trends of actual spending and taxes will be determined by the trend of efficient spending and taxes. The conservative project to reduce the size of government is a struggle against the forces of supply and demand, and as such, not likely ever to succeed.

For the chronic deficit of government spending, in any case, there is a contrary explanation that fits the facts better. It centers on unemployment.

Experience teaches us that when unemployment is high, measures to reduce unemployment often result in government deficits, and there is good reason (from what is called Keynesian economics) to think that deficit spending, per se, is often a necessary means of reducing unemployment. Now, high unemployment is against the interest of both major classes. Clearly it is against the interest of workers, who are the ones who may lose their income prospects because of unemployment. But every unemployed worker is a missed opportunity for profit. Whatever ideology may say to the contrary, government deficit spending that effectively reduces unemployment is in the interest of the capitalist class, as well as the working class. Now, in a society based on class compromise, we must expect that the government will predictably adopt policies that are in the interest of both major classes, and quite simply that is what we have seen. Chronically recurring periods of high unemployment, with the predictable response of governments that act in the interests of the two major classes, have given rise to chronically increasing government indebtedness. A consistent tendency to rising deficits would be a symptom of secular stagnation.

But this needs to be qualified in two ways. First, government debt can become unsustainable, especially when it must be raised at a high interest rate. In some cases we have seen a self-destructive process in which the risk of default leads to rising interest rates that increase the probability that the debt will become unsustainable, and so generate further increases in interest rates. Second, while taxes can be raised, the inefficient excess burden of taxes probably increases more than proportionally. This excess burden could be substantially reduced by a shift to taxes that do not have an excess burden – principally green taxes – but there would certainly be limits to that. In any case taxes seem quite high in many European countries (to the extent that the taxes are paid at all). Rising excess burdens of taxation presumably would set some upper limit to the efficient increase of government spending on the whole.

We have seen that (1) stagnation of wage growth removes one major motive of the working class in entering into class compromise, (2) the cost of assurance of access to medical care and of providing pensions for income from old age, two other benefits the working class receives from class compromise, show chronic tendencies to increase, and (3) government deficits and indebtedness, which arise because of policies to reduce unemployment in the interests of both classes, could at some point become unsustainable. Moreover, (4) in at least some important mature capitalist countries, major institutional bases of the power of the working class, unions and consumer cooperatives, have lost much of the strategic advantage they once had. It seems there are grounds enough to suppose that the forms taken by class compromise in the twentieth century may be infeasible at some point in the twenty-first century. And even if that time is not near at hand, capitalists may see the long-term commitments of their investments as being endangered by the collapse of class compromise, and so begin to withhold investment, as in Lancaster's model.[8] This could bring the crisis forward in time to the near future, or perhaps even the recent past.

In summary, it is not at all clear that the class compromise of the twentieth century can be sustained for the indefinite future. The liberal, democratic market economies of the twentieth century are quite extraordinary in the longer history of the world. It seems that they have emerged because of a balance of power between two powerful classes, each of which owes its power not to its command of weapons but to its economic role. If this balance of power cannot be sustained, whether because of secular stagnation or of the rising cost of public services or for a complex of other reasons, what can we anticipate? Can liberal and democratic laws be sustained, if class compromise cannot?

Socialism

Schumpeter wrote (1928):[3] "Capitalism, . . . will be changed . . . into an order of things which it will be merely matter of taste and terminology to call Socialism or not." Like capitalism and democracy, socialism is a word whose definition is itself a matter of dispute. In this discussion I will follow Nobel Laureate Sir Arthur Lewis[13] in defining a socialist society as one without class divisions, that is, a society in which every person is a member of the same class. The only such possibility is that every person is a member of the working class, a society from which a separate employing class has been eliminated. In practice, realistically, we should probably modify that to say that a capitalist class, if one existed, would be a residual one without relevance to power relations or to key economic trends. In a country like the United States, for example, capitalist agriculture and a similar sector of small business could probably coexist with a working class dominance of political power and the determination of economic trends that would constitute, for all practical purposes, a socialist society. If class compromise cannot be sustained, this would seem to be one of its possible successors.

But could such a society – in which the working class is the only important class – be democratic? The answer seems to be yes. This chapter has argued that democracy arises naturally from capitalism, but it does not follow that democracy requires capitalism. Indeed, the contrary is so. Recall the chain of argument that supported the claim that democracy arises from capitalism. It began with the observation that the capitalist class is not literally a ruling class. The same is no less true of the working class. Thus, the working class, once it is the dominant class, faces the same dilemma that gave rise to democracy on the emergence of mature capitalism: in the absence of a literal ruling class, who shall govern? The options would seem to be the same: strong-man government or representative government. This chapter argued before that strong-man government is more costly to the capitalist class than is class compromise. If we may take our lesson from the Soviet Union, the case is still worse from the perspective of a working class government – strong-man government is fatal, in that it gives rise to a new class(3) that ultimately constitutes itself as a capitalist class.

It follows that, as with capitalism, a sustainable socialism would rely on democratic governance. To the extent that class compromise becomes infeasible, we cannot exclude the possibility that such a system might emerge.

Technocracy

Some recent governments in Italy and Greece have been called "technocratic," in that they have been appointed, not on account of electoral support, but on account of their supposed technical expertise in economics.

"Technocracy" originally referred to a political movement(14) in the United States and Canada based on Thorstein Veblen's proposal of government by "a Soviet of Engineers." This movement anticipated a nonviolent revolution in which a passive collapse of capitalism would lead the elected government to call on the expertise of a Soviet of Engineers (or of the leader of the technocratic movement) to reorganize the economy and society along lines more in accordance with the imperatives of scientific knowledge and technology. By extension, we might describe any government chosen for its supposed technical expertise as technocratic, and this seems to be the current sense of the term. If it were to be more than an emergency arrangement within an ongoing parliamentary democracy, a technocracy would probably have to be constituted as a self-coopting committee or "Soviet" that would coopt new members on the basis of their perceived technical competence. This is the alternative to democratic capitalism that we now explore. There seem to be, in fact, two distinct possibilities. The technocracy might function as a "committee for managing the common affairs of the whole bourgeoisie," that is, as an alternative to class compromise. On the other hand, it might function as the administration of a revised class compromise. These possibilities will be given some separate consideration. However, some further comments seem to be needed on technocracy in general.

The idea of technocracy presupposes that there is some body of technical "knowledge" that, if applied in government, can resolve major problems and give rise to a better public policy result; but that this knowledge cannot be applied because of the inefficient mechanisms of parliamentary democracy. In practice, the technocracy can only be recruited on the basis of what the recruiters *perceive* as knowledge. In the case of the recent technocratic governments in Europe, the knowledge required seems to be the knowledge that government austerity can eliminate government deficits. But this knowledge goes against a great deal of experience that teaches us that austerity leads to worsened recessions that reduce the tax base and thus increase the deficits. One of the dangers of technocracy in general would seem to be that the "knowledge" required of technocrats could in fact lead systematically to bad policy decisions – be "false" in that sense – and in such a case, disastrous failure could be the result.

On the other side, there does seem to be a real problem in introducing reliable expert knowledge into government decisions, in particular, for example, in the design of the tax system.(15) In what follows in this subsection we suspend our disbelief and assume that the technocracy is chosen from those with "true" knowledge, that is, on beliefs that will not systematically lead to self-destructive policy decisions but will improve them by comparison with the decisions of the elected legislatures.

Once again we recall a key point for this chapter: the capitalist class, unlike the feudal landlord class or the citizens in arms of a classical city, is not a literal ruling class. Thus, for a capitalist society, the problem is: who will govern? The two alternatives that have widely been tried are Bonaparte-like strong-man government and electoral, representative government, which in turn entails class compromise; and the latter of these has been the more successful from the point of view of the interests of the capitalist class. Can it be that technocracy provides a third alternative?

In principle the answer seems to be yes. Government by a self-coopting corps of experts might serve the interests of the capitalist class, supposing that the experts possess knowledge of the *actual* interests of the capitalist class and the *actual* means by which these interests can be advanced. These experts would themselves be paid, presumably handsomely, from tax revenues as public employees. But what if they should deviate from the interests of the capitalist class? To subject the decisions of the technocracy to review by some electoral body would bring back the mechanism of electoral democracy and class compromise. Probably, however, the interests of the technocrats could be tied to those of the capitalist class by some other mechanism.

However, while Lancaster may have exaggerated the power of the working class to seize control of production, the working class will retain a capability of disturbing social peace. The removal of advantages the working class has enjoyed in the period of class compromise must be expected to generate such disturbances to social peace. Thus, the technocracy would face in the twenty-first century the choice that capitalism faced in the twentieth: concession or repression. Repression proved the costlier choice in the twentieth century and there is no reason to suppose that it will be otherwise in the twenty-first. Moreover, repression would endanger the technocracy itself, since the means of repression, the military, is itself directed by a hierarchy based on knowledge and experience, a hierarchy that would inevitably take the central position in a repressive technocracy. Thus, a repressive technocracy would rapidly degenerate into strong-man government, to the cost of the capitalist class. Therefore, the technocracy, with its reliable knowledge of the *actual* interests of the capitalist class, would reconstruct class compromise on some new terms that we cannot anticipate.

By way of interim summary, then, it seems most probable that a viable technocracy would not be a substitute for, but an adjunct to, class compromise. It might also be "an order of things an order of things which it will be merely matter of taste and terminology to call Socialism or not" (Schumpter, 1928).[(3)]

CHAPTER SUMMARY

Beginning with a revisit to social contract ideas expressed in terms of non-cooperative game theory, we learn that a cooperative solution to the game of state formation will also be a Nash equilibrium of the corresponding non-cooperative game. However, there are many Nash equilibria of such a game, so we may not be confident that the cooperative solution will be realized. Moreover, this sort of model relies excessively on a representative agent approach, and, in particular, ignores issues of class. Conversely, what we call democracy has historically emerged with the maturation of capitalism. What we seem to observe in successful, mature capitalist economies is a class compromise between the two most powerful classes in a mature capitalist system, with representative government and universal suffrage as one of the terms of the compromise. Such a compromise is a cooperative formation in itself and seems open to pluralistic compromise among factions other than classes. It is, however, an open question whether this twentieth-century class compromise might be sustainable in the twenty-first, and if not, what might succeed it.

CONCLUDING SUMMARY

People can gain a great deal by working together cooperatively. Probably the most important of these gains arise from production, where complex combination labor is a necessary condition for the high productivity of labor characteristic of modern society. For this understanding of production we must return to the economic ideas of Adam Smith and some other classical economists, the Austrian economists, and a synthesis due to the American economist Richard Ely – all largely lost to the economics of the twentieth century. As Smith understood, this complex combination of labor could not be sustained without extensive reliance on exchange. Exchange presents opportunities to benefit from cooperation whenever people have different tastes or endowments, and in particular complex combination of labor generates quite different endowments through specialization in production. Finally, sharing of risk provides further opportunities for mutual gain through cooperation. For our understanding of these opportunities we are indebted to twentieth-century developments in economics and statistics, and to the practice of insurance.

To realize these opportunities, we must coordinate our actions. In the terms of game theory, non-cooperative equilibria will often fail to do this. If information were free, this probably would be irrelevant: people would simply act cooperatively whenever the opportunity for mutual benefit should present itself. However, information is not free but is a high order good (in Austrian terms) that usually must be produced at some cost. Moreover, cooperative action often requires more information or more costly information than non-cooperative action. This leads to a range of consequences familiar from the economic writing of the second half of the twentieth century, from bounded rationality to the design of social mechanisms to reconcile non-cooperative decision-making to cooperative objectives. Prominent among these is organization.

Among the biggest organizations we observe are the national states, another distinctive characteristic of modern societies. If we can understand them as expressions of cooperative action of the whole population, then many of the characteristic activities of states seem predictable – including, in particular, bargaining and compromise. The reliance on a market economy, to the greatest extent possible, appears as a way of economizing on costly information by reconciling non-cooperative decision-making with some cooperative objectives. But this reconciliation is incomplete, and its incompleteness leads to some familiar patterns of government intervention in the economy. The mixed economy emerges predictably in a political economy founded on cooperative principles of efficient compromise. But it is not the only possibility – certainly not the worst, and hopefully not the best.

The modern state, with its electoral institutions and mixed market economy, is associated with mature capitalism. Mature capitalism is unique among historic class societies in that power is divided between two major classes, the bourgeoisie and the working class, both of which derive their power from their economic role rather than from possession of the means of violence. In the circumstances of most mature capitalist countries, in most of the twentieth century, these two classes had some opposed interests but also enough interest in common that class compromise was viable. These circumstances seem to create the possibility of a cooperative polity, a democratic capitalist state as an expression of the compromise between the classes. But as time passes, circumstances change, and the interests of classes change with the circumstances. It is an open question, therefore, whether class compromise can be sustained.

SOURCES AND READING

(1) The ideas of Hobbes, Locke, and Rousseau are widely available and widely discussed. This chapter relies on online sources for Thomas Hobbes (1651), "Leviathan," McMaster University Department of Economics, available at http://socserv.mcmaster.ca/econ/ugcm/3ll3/hobbes/Leviathan.pdf, as of 22 January 2013; on John Locke (1689), "Two treatises of government," McMaster University Department of Economics, at http://socserv2.mcmaster.ca/econ/ugcm/3ll3/locke/government.pdf, also as of 22 January 2013; on Jean-Jacques Rousseau (1762), "The social contract, or principles of political right," translated by G.D.H. Cole, The Constitution Society, available at http://www.constitution.org/jjr/ineq_04.htm, as of 15 January 2013. In normative social contract theory, the work of John Rawls was particularly influential in the twentieth century. See John Rawls (1971), *A Theory of Justice*, Cambridge, MA: Belknap Press. On strong equilibrium see Robert J. Aumann (1959), "Acceptable points in general cooperative n-person games," *Contributions to the Theory of Games, Vol. IV, Annals of Mathematics Studies, No. 40*, Princeton, NJ: Princeton University Press, pp. 287–324.

(2) For the ideas taken from Marx and Engels see Karl Marx and Frederick Engels (1848), "Manifesto of the Communist Party," Marxists.org, accessed 29 August 2009 at http://www.marxists.org/archive/marx/works/1848/communist-manifesto/; Karl Marx and Friedrich Engels (1971), *Capital*, Marxists.org, vol. 1, http://www.marxists.org/archive/marx/works/1867-c1/; vol. 2 at http://www.marxists.org/archive/marx/

works/1885-c2/index.htm; vol. 3, at http://www.marxists.org/archive/marx/works/1894-c3/, all accessed 22 January 2013; and Frederick Engels (1895), "Introduction to Karl Marx's *The Class Struggles in France 1848 to 1850*," accessed 30 August 2012 at http://www.marxists.org/archive/marx/works/1895/03/06.htm. (3) This chapter relies also on Joseph A. Schumpeter (1950), op. cit., Chapter 9 of this volume, note 13. See pp. 73–4 on the bourgeoisie. Schumpeter is, of course, critical to the point of sarcasm of Marx's theory of classes, but nevertheless "we should remember what sort of arguments it replaced . . . the Marxian theory of social classes . . . moves into a more favorable light as soon as we bear this in mind . . . even if we tone it down to the proposition that historical events may often be interpreted in terms of class interests and class attitudes and that existing class structures are always an important factor in historical interpretation, enough remains to entitle us to speak of a conception nearly as valuable as was the economic interpretation of history itself," op. cit., pp. 13–14. This seems a sufficient common ground. Note also Joseph A. Schumpeter (1928), "The instability of capitalism," *The Economic Journal*, **38**(151) (Sept.), 361–86. See pp. 385–6. I am indebted to Martin Gaidarov, an undergraduate student in my senior seminar in winter quarter, 2012, for bringing this paper to my attention. The paper is not, as one might suppose, a preliminary statement of the theme of *Capitalism, Socialism and Democracy*, 1942, but rather concentrates mainly on arguments that markets per se are unstable, which Schumpeter rejects; but he comments very briefly at the close of the paper along the lines indicated. The ideas of Sir Arthur Lewis (1969), op. cit., Chapter 8 of this volume, note 18, have also influenced this discussion. On the concept of a new class in the Soviet-type regimes see Milovan Djilas (1957), *The New Class*, London: Thames and Hudson. (4) I follow Samir Amin (1989), *Eurocentrism*, New York: Monthly Review Press, in identifying many pre-feudal societies, including the Roman Republic, as tributary. See, for example, p. 114. Despite characterizing the transition from the Roman state to a feudal one as an advance, Amin does not distinguish feudal from tributary societies. While I regard this as an error, it would reinforce the point made in the text. (5) Compare Schumpeter, op. cit., Chapter 9 of this volume, note 13 (1950, p. 285). (6) Compare Schumpeter, op. cit., Chapter 9 of this volume, note 13 (1950, p. 128). This would seem to contradict the Leninist argument that war, by stimulating aggregate demand and conquering new dependent markets, increases the profits and so favors the interest of the capitalist class. So be it. The experience of the twentieth century establishes that the dependent markets founded on imperialism simply are not profitable markets. Note also Kevin Neuhouser (1993), "Foundations of class compromise: A theoretical basis for understanding diverse patterns of regime outcomes," *Sociological Theory*, **11**(1) (Mar.), 96–116, on the relation of strong-man government to market processes.

(7) From Eduard Bernstein (1899), "Evolutionary socialism," Marxists.org, at http://www.marxists.org/reference/archive/bernstein/works/1899/evsoc/index.htm, pp. 122–3 as of 22 January 2013: "Universal suffrage in Germany could serve Bismarck temporarily as a tool, but finally it compelled Bismarck to serve it as a tool . . . universal suffrage is only a part of democracy, although a part which in time must draw the other parts after it as the magnet attracts to itself the scattered portions of iron." From p. 109: "the co-operative stores . . . in the course of time have really proved to be an economic power – i.e., as an organism fit to perform its work and capable of a high degree of development." From p. 119: "The trade unions are the democratic element in industry." In each case Bernstein contrasts these observations with the expectations of revolutionary socialists. From p. 183: "Constitutional legislation . . . Its path is usually that of compromise."

(8) The concept of class compromise will be familiar to those who have read the work of Adam Przeworski. See in particular Adam Przeworski and Michael Wallerstein (1982), "The structure of class conflict in democratic capitalist societies," *The American Political Science Review*, **76**(2) (June), 215–38. The model central for Przeworski and Wallerstein is similar to that of Kelvin Lancaster (1973), "The dynamic inefficiency of capitalism," *Journal of Political Economy*, **81**(5), 1092–109, whom they cite, without the longer-term foresight Lancaster assumes and without the consequent breakdown of capitalism that his model predicts. In subsequent writing derived from Przeworski's, the game theoretic discussion is generally quite elementary. Marick F. Masters and John D. Robertson (1988), "Class compromises in industrial democracies," *The American Political Science Review*, **82**(4) (Dec.), 1183–201, treat government as an autonomous "player in the game" and describe class compromise as a "tripartite coalition," seeing growing government expenditure as one element in the compromise. John D. Robertson (1990), "Transaction-cost economics and cross-national patterns of industrial conflict: A comparative institutional analysis," *American Journal of Political Science*, **34**(1) (Feb.), 153–89 uses transaction cost economics to refine a discussion of the forms class compromise may take. While his statistical studies should be read with caution, he offers some hints as to why electoral democracy is always a part of it. Desmond A. King and Mark Wickham-Jones (1990), "Social democracy and rational workers," *British Journal of Political Science*, **20**(3), 387–413, draw negative conclusions about class compromise on somewhat contradictory premises. Before p. 408, they seem to consider class compromise exclusively in terms of concessions by

the working class, which they justify with a two-by-two non-cooperative game. They then briefly consider a "cooperative solution" but assert, in this context, that the capitalist class would have negligible bargaining power. Erik Olin Wright (2000), "Working-class power, capitalist-class interests, and class compromise," *American Journal of Sociology*, **105**(4) (Jan.), 957–1002 elaborates Przeworski's model with a broader menu of benefits and costs of compromise and a discussion of conditions that may shift the class compromise, which he identifies (p. 976) with a Prisoner's Dilemma played repetitively. None of these cite Lancaster's paper, nor do any make reference to the relation of capitalism to innovation or to the growth or stagnation of innovation opportunities.

(9) Of the many studies of rational action in politics this discussion draws especially on two: William H. Riker (1962), *The Theory of Political Coalitions*, New Haven, CT: Yale University Press, and Mancur Olson (1971), *The Logic of Collective Action*, Cambridge, MA: Harvard University Press. (10) The concept of externality used here is, of course, due to Arthur C. Pigou (1920), op. cit., Chapter 8 of this volume, note 2. On technological trends and externalities, see Roger A. McCain (1978), "Endogenous bias in technical progress and environmental policy," *American Economic* Review, **68**(4) (Sep.), 538–46. The methods used in this study have been largely abandoned in recent economic research, but they capture one thing that most more recent research on the direction of technical progress misses: researchers, like other economic decision-makers, face limited possibility frontiers.

(11) *The Cost Disease: Why Computers Get Cheaper and Health Care Doesn't* by William J. Baumol, David de Ferranti, Monte Malach and Ariel Pablos-Méndez (2012), New Haven, CT: Yale University Press. The argument is (in terms used in neoclassical economics) that, for efficient provision marginal benefit should be equal to marginal cost. It is always assumed – and probably usually correctly – that marginal benefit decreases as the level of provision increases. The rate at which the marginal benefit decreases as provision increases is measured by the "elasticity of demand," and while this concept is derived from market economics, it can be applied also to public services. (For these concepts see Chapter 8 of this volume, especially the footnote in the first section.) Baumol's "cost disease" will be observed for any product, whether privately or publically supplied, for which the elasticity of demand is less than 1 and labor-saving technical progress is less than the average rate in the economy as a whole. In the case of medicine, it seems that technical progress is not labor saving but quality enhancing. It is also usually the case that the marginal benefit will increase as total income increases. We can compute an elasticity of demand for the service with respect to income, and if it is greater than 1, this in itself will

cause an efficient level of spending on the service to increase faster than income. John Kenneth Galbraith (1958), *The Affluent Society*, New York: New American Library, explained the increasing level of government spending in those terms. Note also Roger A. McCain (1974), "Induced bias in technical innovation including product innovation in a model of economic growth," *Economic Journal*, **84**(336) (Dec.), 959–66 on quality-enhancing technological progress.

(12) This idea seems to have been returned to the awareness of economists by Robert J. Barro (1974), "Are government bonds net wealth?" *Journal of Political Economy*, **82**(6) (Nov.–Dec.), 1095–117. (13) On Lewis' conception of socialism see Chapter 8 of this volume, note 18. (14) Technocracy, Inc. has an office in Ferndale, WA, and a website at http://www.technocracy.org/, last accessed 19 November 2013. For the ideas of Veblen, most relevant to the technocracy movement, see Thorstein Veblen (2001; first published 1921), *The Engineers and the Price System*, Kitchener, ON: Batoche Books. (15) On the problem of designing an efficient system of taxation, see James A. Mirrlees et al. (2011), op. cit., Chapter 8 of this volume, note 3.

Index